An Introduction to Astrophysics with Python

Stars and planets

Online at: https://doi.org/10.1088/2514-3433/ade5f6

AAS Editor in Chief

Ethan Vishniac, Johns Hopkins University, Maryland, USA

About the program:

AAS-IOP Astronomy ebooks is the official book program of the American Astronomical Society (AAS) and aims to share in depth the most fascinating areas of astronomy, astrophysics, solar physics and planetary science. The program includes publications in the following topics:

GALAXIES AND COSMOLOGY

INTERSTELLAR MATTER AND THE LOCAL UNIVERSE

STARS AND STELLAR PHYSICS

EDUCATION, OUTREACH, AND HERITAGE

HIGH-ENERGY PHENOMENA AND FUNDAMENTAL PHYSICS

THE SUN AND THE HELIOSPHERE

THE SOLAR SYSTEM, EXOPLANETS, AND ASTROBIOLOGY

LABORATORY ASTROPHYSICS, INSTRUMENTATION, SOFTWARE, AND DATA

Books in the program range in level from short introductory texts on fast-moving areas, graduate and upper-level undergraduate textbooks, research monographs, and practical handbooks.

For a complete list of published and forthcoming titles, please visit iopscience.org/books/aas.

About the American Astronomical Society

The American Astronomical Society (aas.org), established 1899, is the major organization of professional astronomers in North America. The membership (~7,000) also includes physicists, mathematicians, geologists, engineers, and others whose research interests lie within the broad spectrum of subjects now comprising the contemporary astronomical sciences. The mission of the Society is to enhance and share humanity's scientific understanding of the universe.

An Introduction to Astrophysics with Python

Stars and planets

James Aguirre

*Department of Physics and Astronomy, University of Pennsylvania,
Philadelphia, PA, USA*

IOP Publishing, Bristol, UK

ISBN   978-0-7503-5467-7 (ebook)
ISBN   978-0-7503-5465-3 (print)
ISBN   978-0-7503-5468-4 (myPrint)
ISBN   978-0-7503-5466-0 (mobi)

DOI   10.1088/2514-3433/ade5f6

Version: 20251201

AAS–IOP Astronomy
ISSN 2514-3433 (online)
ISSN 2515-141X (print)

British Library Cataloguing-in-Publication Data: A catalogue record for this book is available from the British Library.

Published by IOP Publishing, wholly owned by The Institute of Physics, London

IOP Publishing, No.2 The Distillery, Glassfields, Avon Street, Bristol, BS2 0GR, UK

US Office: IOP Publishing, Inc., 190 North Independence Mall West, Suite 601, Philadelphia, PA 19106, USA

To my students, who pushed me to make this better

To my family, for their patience and support

And in memory of Ken, who inspired me

Contents

Acknowledgments

I would like to thank Masao Sako and Robyn Sanderson for their comments and help with manuscript, and Saul Kohn for being an amazing TA the first time we tried this.

Introduction

What This Book Is and Isn't

This book is intended to provide a modern introduction to some key ideas in astrophysics with an emphasis on the underlying *physics* in the astrophysics. However, it is, by design, different from most introductory texts on astrophysics. It is *not* a comprehensive introduction to the vast panoply of modern astrophysics, or an extensive reference text, for which there are many good options available. Rather, this book is intended to foreground the key concepts and ideas while only relaying as much factual information as is absolutely essential, and to move the reader as rapidly as possible to applying the ideas to solving reasonably sophisticated problems to better get a sense of modern practice in the physical sciences. As the title indicates, these problems focus on numerical calculation and Python as a programming language, due to Python's current ubiquitous use in astrophysics.

The book reflects my own experience in teaching a course using this material for many years at the University of Pennsylvania ("Penn"). The course started out as a traditional lecture and has evolved into its present form with interactive in-class exercises and numerical work over the past decade. The course at Penn is offered as a second-year course to physical science and engineering majors who have had introductory calculus and mechanics but who have generally not had more advanced physics courses, including electrodynamics, thermodynamics and statistical mechanics, and quantum mechanics. Since astrophysics requires some concepts from all of those fields, these need to be (gently) introduced along with the astrophysical setting as well as the numerical aspects.

My goal in writing this is to address the needs of both the student and the instructor for a course like this. Specifically, I have attempted to provide clear and concise explanations to convey the information to the reader and written exercises (which are an essential part of the text!) that will hopefully provoke a useful interaction between the student and instructor, or between the reader and text. In particular, I want to show more how modern practice uses numerical calculations to address questions that are beyond what can be addressed analytically, and to emphasize that most modern physical science involves building sophisticated mathematical *models* of physical phenomena that attempt to explain both how a given process works and to provide quantitative *predictions* of observable phenomena that can be compared to experimental or observational measurements. This model building is generally complicated enough that it is not possible to write down solutions for the mathematical equations describing it in any simple form, and hence we need to use a numerical approach. Indeed, in our first major example—the relatively simple problem of two bodies interacting only via gravity problem—famously does not have an analytic closed-form solution for the position of the bodies as a function of time, and thus requires a numerical method to solve the implicit equations for this position. In addition, in order to understand how a complicated physical model actually works, we generally need to be able to vary the input parameters and evaluate the model's behavior over a wide range of values, which necessitates a lot of computation, which we wish to push off on the computer.

In some ways, a book structured like this necessarily values method and process over factual knowledge. Having pared the background information down to the bare minimum does lead to the risk that the reader's takeaway will be that this is all there is to know. I have tried to counteract this as I explain below. There is also the fact that, to be tractable, all the models presented here are in a very real sense "wrong", and can depart in important ways from actual facts and observations. I have tried to emphasize the importance of gaining some physical intuition and expectation by simplification, while leaving open the construction of more complex (and hopefully accurate) models as something the reader will want to explore further.

Outline of the Book

The astrophysical content of the book focuses on stars and planets. For the physics content, I have restricted myself to a level accessible to students in their second year in a US college or university, and this results in what may seem at first glance a rather bland set of topics: non-relativistic Newtonian mechanics and gravity and a simplified non-relativistic approach to quantum mechanics to describe the inter- action of light and matter, combined with a bit of statistical physics. Nevertheless, students should be impressed that even with these modest materials, we can still describe a tremendous variety of phenomena of interest, including the motions of a wide range of astrophysical objects, the atmospheres and interiors of planets and stars, and the main methods for detecting and characterizing extrasolar planets. The remaining physics to fully appreciate the rest of modern astrophysics is, of course, both special and general relativity, with all of its application to black holes and cosmology and a host of "exotic" phenomena. Presenting that material is beyond the scope of what can be done here and merits its own book.

The material is presented in three major blocks. The first section is designed as an introduction to Python in the way we will use it here: as a method to do numerical calculations, solve differential equations, and make graphs to interpret results.[1] I focus primarily on an important problem—both historically and practically—and indeed the problem that fundamentally put the physics in astrophysics: the motion of two bodies interacting only under their mutual gravitation. The astrophysical applications of this "Kepler" or "two-body" problem are widespread, from calculating the positions of planets in the solar system, to determining the masses of stars and planets, to finding and characterizing exoplanet systems. While the two- body problem represents an idealization (the gravitational interactions of any system generally involve more than two bodies), it is so useful, and the intuition we will build so powerful that, combined with what we will learn about numerically solving differential equations, it merits taking up the first third of the text. Beginning with Newton's Laws allows the reader to start from a place where they are comfortable as

---

[1] I introduce only as much computer science and "programming" as is necessary to get the job done.

far as the physics content, and I use this to introduce a number of important concepts in Python, including how to do numerical calculations, use units, define functions, and produce plots. We then become more sophisticated in developing the idea of solving differential equations (a key skill in any physical science) and develop the methods for solving two-body differential equations numerically. This allows us to study basically every part of this interaction: the position and velocity versus time, the energy and angular momentum, and dependence of all these things on the initial conditions. There are two capstone exercises associated with this section: a fully worked-out numerical calculation of orbits under various initial conditions and a qualitative exploration of how these rules govern orbital dynamics of a satellite.

The middle section changes focus to properties of light and its interaction with matter, as despite exciting advances in gravitational wave and neutrino detection, light remains the primary means by which we obtain information from the cosmos, and its interaction with matter the primary way we have of inferring the properties of the objects that emitted it or interacted with it on its journey to us. This section is considerably lighter on numerical calculations, but develops a wide range of physical concepts that are likely to be new to students. The capstone exercises here use the Saha equation to understand the ionization behavior of hydrogen in a stellar atmosphere and velocity measurement from Doppler-shifted lines in stellar atmospheres to detect exoplanets.

The third part brings together the ideas in the first two parts and is the most rich in applications. While continuing to introduce new material, almost every chapter also attempts to synthesize multiple lines of reasoning to build models for the surface temperatures of planets and the behavior of the atmospheres and interiors of stars and planets. Our introduction to the stellar main sequence is a brief one, but the capstone exercises expand the key ideas by constructing a model for absorption lines in stellar atmospheres and how the line strength varies along the main sequence, as well as indicating how to construct a model of the stellar interior, and exploring its implications with a model of the Sun.

At the end of each major part, I include a review of the major concepts and takeaways. I also provide a small survey of additional topics to emphasize that the material presented is only a small sample of what is known, and indicate some additional directions the students could take things. In particular, there is a great deal more to say about gravitational dynamics, radiative transfer, and post-main-sequence evolution of stars, to name just a few topics.

The parts of the book build on each other sequentially and are intended to be read in order. I have tried to explicitly indicate where connections between ideas and equations occur as much as possible. In working through the exercises and problems, the reader should definitely refer back to previous material!

This Book and Active Learning

The advent of the Internet sparked a revolution in the ability of anyone to find and use information. This information is widely and freely available at the touch of a keyboard, and in great detail, to anyone who cares to look. This situation has become even more dynamic, as at the time of writing, the use and capabilities of generative AI are exploding, and it is not at all clear what the implications will be for teaching and learning. So what is the reason for reading this book, or taking a university course, if the information is already all out there?

In part, I would answer that knowledge is not something that exists externally to human minds and can simply be stored in an algorithm or on a disk. Scientific learning must be practiced, wrestled with and internalized to be truly learned. Once learned, it can be synthesized and put to all sorts of new uses and extended in ways that will continue to surprise. Given the rapid development of generative AI, almost anything said here is likely to be out-of-date almost immediately. However, we have to believe that training actual human minds to understand the universe is still important, no matter how successfully the machines are able to do it.

With the goal of continuing to educate human minds, we note that a considerable body of education research has now shown that the traditional lecture method of presentation is terribly inefficient at actually producing reliable recall of material, much less real understanding of it. Even the venerable problem set, a staple of physics classes for generations, doesn't always produce these results. This was brought home to me by the (very bright!) student who commented at the end of the semester: "The one issue with this course was that I and others could complete the homework without having any idea what was actually going on." Clearly, this was not the desired outcome!

To address this, I began adopting an approach called Structured, Active, In-class Learning (SAIL).[2] I have used this "SAIL" approach with variations on the material in this book four times over the last decade. The idea was that each class session would be organized around an extended exercise, here included as the final section of each chapter with each chapter corresponding roughly to one class period. I have deliberately chosen to present only as much material as I have successfully presented in a one-semester course.[3] As noted above, this is justified by the need to introduce both computational and physical concepts in addition to astrophysics. The idea is that the material in the chapter is read and some reading questions completed prior to the class meeting. The instructor can then review important points or mis-conceptions that arise from the reading questions. The bulk of the class time is spent on the exercises, which the students work through in small groups (I have found that groups of 3–4 students are optimal) and which are overseen by the instructor and teaching assistants. This discussion—which cannot be reproduced in a book—is often the most productive part of the learning experience. If you are reading this book on your own, you will certainly learn something from the text itself, but most of the learning will take place in working through the exercises.

In writing the exercises, I strove to structure them in a way suitable to the classroom environment. First, some background is required in order to understand what questions about the Universe might be interesting. Second, this background

---

[2] This method goes by different names at different institutions, and emphasizes different approaches in the classroom, but all share the common feature of interactive rather than passive learning.

[3] At Penn, the semester lasts 14 weeks with approximately 28 class meetings.

attempts to lead naturally into a specific question one might ask. Generally, such a question, while interesting, is designed to be sufficiently difficult that students will not immediately know how to answer it. This leads to the third step, a bit of brainstorming and an outline of how to tackle the problem. It is important to do this step and not just immediately move into a prescribed set up of steps for solving the problem, lest the students not understand the underlying motivation and see the problem-solving process purely as a recipe. The lion's share of the time will, of course, be spent on actually solving the problem once the line of attack is understood, but then, once the problem has been mostly solved, it is important to take stock and assess what has been learned. In some cases, this will be specific skills, but in general we want to see what appreciation for the original question has been gained by working through the problem. This is the point at which one should ask "What did you learn? What was surprising? What's the big takeaway?"

This process is intended to teach "practice appropriate to the field"; that is, not just explain the results, but explain the way practicing astrophysics ask questions and think about and solve problems. This includes technical aspects like parts of physics, and also techniques like numerical problem solving. Another is engaging in activities that are like what professionals in the field do: sophisticated, multi-part problem solving; discussion of results and methods with peers; and the interpretation of results. It also addresses the fundamental problem in teaching physics that the student's quote notes: the lack of connection between theoretical constructs and mathematics on the one hand and the physical world (the real objects and phenomena) they are supposed to describe on the other. My hope is that this book will help you develop the ability to understand the way in which such mathematical explanations actually do explain the features of the observed universe.

The Role of the Computer

An important element of my approach in this book is to integrate the computer into the material and demonstrate how to think about scientific problems numerically, in addition to the traditional analytical approach, and to make numerical calculations an essential component. This attempts to address what I perceive to be a continuing gap between the actual practice of physics and the present education of our physics majors. The problems in this book use both real and synthetic data, and quantitative interpretation of its meaning. The numerical topics addressed included numerical differentiation and integration, root finding, visualizing and plotting data, and computing simple solutions to differential equations. The approach taken here is to start with simple programming features and slowly build up complexity by examples. `Python` is a fairly user-friendly language with a wide base of support online.

In spite of the centrality of the code, it is important for the reader to keep in mind that the computer is intended as an *aid* in solving problems; if you are finding it to be a hindrance, it is good to step back and think about how one would go about solving the problem if a computer were not involved.

Other Features of the Book

There are two other features of this book that merit a brief comment: the choice of how to present mathematical derivations and the use of figures. In my experience, the value of presenting derivations, either on a chalkboard or as long sections of text to be read, is decidedly mixed. On the one hand, it is highly desirable that students learn that the "boxed equations" are not handed down from on high, but come about via a process of physical reasoning, mathematical manipulation, and a clear view of an appropriate physical interpretation of the mathematical symbols. In this view, the presentation of a derivation highlights both the method by which the equations are arrived at and the numerous assumptions and approximations on which every equation relies. Moreover, correctly deriving new equations from known ones is a skill that should absolutely be developed in a student of the physical sciences, and helping students develop this skill is absolutely of pedagogical value. On the other hand, the amount of detail required to make the derivations actually clear without requiring inordinate work on the part of the reader requires a considerable amount of space and can dissipate the logical flow of a presentation, causing one to lose the forest for the trees. It is also a skill best learned through practice rather than example. Here, I have tried to have it both ways: I generally include the derivations of important equations to show the method and the reasoning where I think it is helpful to the discussion and a student's understanding but have foregone them when the derivation leads too far afield, requires concepts that are unlikely to be known to the reader, or are not going to be used again in this book, or in cases where the derivation is sufficiently lengthy that it would detract from the overall presentation. (In one particular case (blackbody radiation) I have relegated a more extensive derivation to Appendix A.) This has meant that some equations, like the Saha equation, are presented essentially by fiat, which is undesirable. Hopefully in cases where the derivation of equations is unsatisfactory to the reader or an instructor, the relevant details can easily be found and filled in from other sources.

An important final feature of this book is the figures. I have striven to use real data or calculations in producing the figures, rather than relying on schematics or cartoons. As one of the skills a reader should learn from this book is the production of high-quality, clear figures, many of the exercises involve producing figures which are highly beneficial to a full understanding of the material. All figures in the text (which are not astronomical images in the public domain) are ones which are possible to generate using the knowledge of astrophysics and `Python` contained in the book. Indeed, the `Python` code necessary is publicly available.

Author biography

James Aguirre

James Aguirre has been a professor at the University of Pennsylvania since 2008. He received undergraduate degrees from Georgia Tech in Physics and Applied Mathemathics, and received his PhD in Physics from the University of Chicago. He was a postdoc and an National Radio Astronomy Observatory (NRAO) Jansky Fellow at the University of Colorado, Boulder. He is broadly interested in the formation and evolution of galaxies and the interplay between galaxies and the large-scale structure of the universe. His research focuses on the use of radio and far-infrared wavelength telescopes to study galaxies and he has worked on a variety of experiments, including several using high-altitude balloons to carry telescopes above the Earth's atmosphere.

This is his first book.

Part I

Newton's Gravity and Orbits

An Introduction to Astrophysics with Python
Stars and planets
James Aguirre

Chapter 1

A Review of Newton's Laws and an Introduction to Python

This chapter reviews Newton's three laws of motion, and his law of gravity, which form the foundation for studying the motions of planets and orbiting bodies. We also introduce the `Python` computing language and `Jupyter` notebooks and show how they can be used to make numerical calculations and keep track of units in physics problems using `astropy` units.

1.1 Overview

To begin our study of astrophysics, we'll begin with a little refresher on Newton's Laws (particularly the Second Law), and the SI (a.k.a metric) units associated with quantities like force and momentum. We'll introduce `Python` as a way to do calculations, make simple plots and keep track of units. Then we'll talk about how to think about mathematical functions as both operations a computer can do, and as arrays of numbers the computer can store, and in general how to make calculations less tedious than pencil-and-paper and calculator. In subsequent chapters, the material is presented first and then there is a follow-on exercise; this chapter is different in that the exercise is integrated with the reading.

1.2 Scientific Background

We'll start by recalling that Newton gave three laws of motion, roughly paraphrased as

1. An object in uniform straight-line motion or at rest stays in that state of motion unless an external force is applied.

doi:10.1088/2514-3433/ade5f6ch1

2. A force applied to a single particle of mass m causes an acceleration according to

$$\mathbf{F} = m\mathbf{a} = m\frac{d^2\mathbf{x}}{dt^2} = m\frac{d\mathbf{v}}{dt} = \frac{d(m\mathbf{v})}{dt} = \frac{d\mathbf{p}}{dt} \qquad (1.1)$$

where, in the modern notation, the force $\mathbf{F}$, acceleration $\mathbf{a}$, velocity $\mathbf{v}$, momentum $\mathbf{p}$, and position of the particle $\mathbf{x}$ are three-dimensional vectors that are functions of time t, and the Second Law is equivalent to a differential equation for the position as a function of time of the particle $\mathbf{x}(t)$.

3. If two bodies exert forces on each other, the force exerted by the first body on the second is equal and opposite to the force exerted by the second body on the first.

Newton also gave a specific law for the gravitational force between two objects, which will turn out to be paramount in astrophysics: gravity is only one of two long-range forces, and often a dominant force in astrophysical problems. The form of Newton's Law of Gravitation for the force exerted by particle 1 of mass m_1 on particle 2 of mass m_2, with the particles located at positions $\mathbf{r}_1$ and $\mathbf{r}_2$, is

$$\mathbf{F}_{21} = G\frac{m_1 m_2}{r^2}\hat{\mathbf{r}} \qquad (1.2)$$

where $\mathbf{r} = \mathbf{r}_1 - \mathbf{r}_2$ and $\mathbf{F}_{12} = -\mathbf{F}_{21}$.

For the rest of this chapter, we'll ignore the vector nature of Newton's laws (though we'll come back to it) and just think about how to do calculations with physical quantities at all using the computer.

1.3 Using Python as a Calculator

We'll start by using Python as a glorified scientific calculator.[1] We'll use a nice web-based interface to Python called jupyter notebook. Start this up by clicking on the icon in the start menu or launcher or by typing jupyter notebook at the command line.

You'll need to make sure you are saving your notebook someplace sensible, so when the web browser first opens up, navigate to a folder where you'd like to save the results (or create a new one) and then select New → Python 3. Then rename the notebook from "Untitled" to something you'll remember (like "IntroductionToPython"). Note that the notebook is saved with a .ipynb extension. Verify that you can find where on your computer this was saved.

There is immediately a place to type, called a "cell" and we could start entering code straightaway. But let's go to the menu and select Cell → Cell Type → Markdown, which now makes a cell where we can type ordinary text (or math, if you happen to know LaTeX). You should use these markdown cells to add

---

[1] It can also be a graphing calculator, which we'll come to shortly.

annotations to your notebook, and in particular to answer the questions enumerated below. In the first cell, enter the names of the members of your group, and then type Shift-Enter (together).

In the next cell, let's now do some calculations: type

```
3+3
```

and then run the cell by using the right arrow or typing Shift-Enter. (Note that typing Enter just moves to a new line.) Does the result look reasonable? How is this different from

```
3+3.
```

(Note the extra dot. Its meaning is subtle; can you work it out?) What about

```
3*3
```

and

```
3/3
```

and

```
4/3
```

Great, so Python knows about arithmetic.[2] It also has some rules for real (really, rational) numbers and integers.

Another useful thing is to be able to use scientific notation. How do we enter a number like 3×10^{12}? There are actually several ways to do this. We can type

```
3*10**12
```

Notice how we're expressing exponentiation. We could also type

```
3e12
```

Notice that this has the same meaning. Also, notice that "e" here is *not* the Euler constant e, but a stand-in for "times ten to the power following the e". This can be kind of confusing at first, but the compact notation is useful.

So, what about trig functions?

Question 1. What happens when you enter `sin(1.)`? Look carefully at the output and try to explain what Python is telling you. Explain this is in a Markdown cell.

---

[2] Hopefully this makes clear what symbols Python is using for addition, multiplication and division. What about subtraction?

Well, that's really going to be a problem for a(n) (astro)physics course! But we have learned something important: `Python` tries to tell us what's wrong, and in general you should start at the last line of gobbledy-gook to start decoding the problem.

`Python` by itself has very limited capability for doing scientific computing, but fortunately, many people have written additional software, called "modules"[3] for it. For now, we will issue the cryptic command

```
import numpy as np
```

and now we can do

```
np.sin(1.)
```

Basically, this tells `Python` that it should look for the sin function in the numerical `Python` (`numpy`) module, which we're going to call `np` for short.

Notice that we can also make assignments of the output to variables to save for later, so that

```
a = np.sin(np.radians(45))
```

As usual, the argument of sin needs to be in radians, but `numpy` provides a convenient method for doing the conversion. Also notice that "=" doesn't mean a logical test for equality, but rather "assign the value on the right hand side the name on the left."[4] Notice that running this cell doesn't seem to do anything, but if I now say

```
print(a)
```

something interesting appears.

Question 2. What did you expect for the answer here? Notice that `numpy` is directly evaluating the numerical value, and not expressing it as you might, with square roots. Can you figure out how to get `numpy` to evaluate what you know to be the right answer from trig? A helpful thing to know here is that typing `np.` and the "Tab" key will show possible completions for `numpy` functions.

---

[3] Don't worry about what a module actually "is" just yet. For now, think of it as some additional functionality we're going to add to bare `Python`, and that we need to request specifically.

[4] This use of the "=" to mean assignment rather than mathematical equality (which for example, is the same regardless of which side of the = things appear on), is quite old in computer science, and isn't changing now. It may take some getting used to, or you might not find it confusing at all.

1.4 Arrays and (Avoiding) Loops

You may have learned in another course on programming about the all-important for loop. And you may be tempted to use for loops to repeat calculations for many different values, but in Python we can mostly use *arrays* (lists of numbers stored in a single variable) instead: most common functions are defined such that if you feed an array to the function instead of a single value, it will *automatically* loop through the values in the array. It is very rare that you actually have to write a for loop in Python!

For example, I want to calculate values of the cos function for a bunch of different input arguments. Since doing many tedious calculations is what we have computers for, let's consider what's necessary. First, I need an independent variable x that runs over some range (say one period), and a variable to hold $\cos(x)$ for every value of x. These two variables are examples of numpy arrays, a data type defined in the numpy module. We can create them as (type along here):

```
x = np.linspace(0,2*np.pi,num=100)
y = np.cos(x)
```

The function np.linspace creates an array of 100 evenly spaced values of x between 0 and 2π, and y (created by calling the cosine function with x) now holds the values of the cosine for each value of x. The cosine function is an example of a function that will accept either a variable holding a single number or an array, and return something with the same type as its argument. If we did it right, y should now also be an array with 100 elements, in the same order as x. You can verify this, since numpy arrays have a method to tell you their "shape", in this case, how many elements:

```
print(y.shape)
```

You can ask for all the values of y by typing

```
print(y)
```

or you can ask for any specific one by saying

```
print(y[15])
```

which will give the 16th in the list (we number from 0 in Python). The number 15 in the example is called an *index*, an integer that specifies the position of a value in the array. Every value in the array has an index. We can also get a "slice" of the array, which is some subset of it defined by a range or list of indices. The : operator is shorthand for specifying a range; this range includes every value between the first one (inclusive) and the last one (**exclusive**). What does, for example,

```
print(y[15:18])
```

produce?
Now try

```
print(y[15:18:2])
```

and

```
print(y[18:15:-1])
```

and see what you get!
Often we want to find a specific number in a list of numbers; we can use various functions to figure out the index of a specific number.

Question 3. Suppose we want to find the index of the array element that is the largest. We can use np.argmax to do this. At what indices does our $y = \cos x$ array have a maximum?

It would be much more interesting to plot this than to look at the list of numbers, and here we need another module:

```
import matplotlib.pyplot as plt
plt.plot(x,y)
plt.xlabel('x [dimensionless]')
plt.ylabel('y [dimensionless]')
plt.show()
```

The basic logic of plotting is to create the plot, add annotations, additional curves, and then at the end show it with plt.show(). The above should let you see a plot of the cosine.

Question 4. Now calculate $\sin(x)$ for the x values you have defined. Add it to the plot of the cosine (just add another plt.plot line in the same cell). Also calculate $\sin(x)^2 + \cos(x)^2 - 1$ and plot it on the same plot. (If you already know something about Python or other coding languages, you might be tempted to do something here involving a for loop or similar. Resist the urge! You can write the above expression very much like you would in a math course, and it will be just fine. Remember, many functions are defined for arrays as well as individual numbers!) You can raise numbers (or each element of a list of numbers) to a power in numpy using either the np.power function (defined for any power including non-integers) or the ** notation from earlier (defined only for integer powers, but faster than np.power when you can use it). Here, use np.power(np.cos(x),2) to square each element of $\cos(x)$, in order to try it out. Don't forget to label your axes!

1.5 Using `Python` Modules to Help with Units

You've probably had it drilled into you that physical quantities always have a magnitude and a unit.[5] And you've also learned that units obey algebra as well: products and quotients of units make new units, and sums of dissimilar units are nonsense. Further, when taking a sin or exp or ln of something, the argument should be dimensionless.

The structure of `Python` makes it possible to glue numbers and their units together, and to do sensible things with their arithmetic combinations. A very nice implementation of this is done in the module `astropy`, which as the name implies, has lots of nice astronomy-related features and functions.

In a new cell, type

```
from astropy import constants as c
from astropy import units as u
```

Now we have units available to us from the u module, and physical constants (including useful astronomical ones) from c. So now

```
x = 1.*u.meter
```

defines *x* as a "Quantity object" equal to 1 meter (value and unit!). An *object* is another computer science concept. In brief, objects have both data and some defined operations you can do on the data (its *methods*.) The data associated with objects are referred to as its *attributes*, and are not limited to the value of the variable. For example, the shape of a `numpy` array is one of its attributes, which you accessed earlier. Nearly everything in `Python` is, or can be thought of as, an object, including the arrays you already defined—and the modules you loaded!

Similarly, if we define

```
t = 5.*u.second
```

notice that

```
v = x/t
```

defines a new Quantity with the right units (and we had to do nothing!). Now, you can see that this new capability is useful in a number of ways:

- You can keep track of the units in your calculations easily
- You can avoid mistakes with units
- There is even a way to do conversions between units[6]

---

[5] And this book will be no exception!
[6] This is one of the operations, or methods, defined for the Quantity object.

Question 5. Let's see that Point 1.5 is true. What happens if you type x + t? (It will look like Python got very angry, but remember to scroll to the last message.) Briefly explain why this makes sense.

Objects are really what Python is all about, and while I won't go into detail, you can start to see how they are useful by noting that the thing we called v isn't just a number, but it has units (another example of an attribute), and moreover, you can *do* things with it. So regarding Point 3, Quantity objects come with a useful method called to, which allows conversions:

```
v.to(u.kilometer/u.second)
```

Question 6. What happens when you type x.to(u.second)? What about x.to (u.imperial.mile)?Again, explain why this makes sense.

Wow, Python is going to save us tons of time and errors!

Now, we can also use those physical constants, all of which are Quantity objects, to do something like calculate a force:

```
m = 1.*u.kilogram
r = 1.*u.meter
F = c.G * np.power(m,2)/np.power(r,2)
```

Notice that we've indicated the function of raising to a power with a somewhat long notation.

Question 7. What is the value of F above in Newtons?

A subtle point is in order about the to method: if you type the last line at the command prompt, you will get back *F* converted to Newtons. However, if you ask about *F* again, you will find its units are still reported as kg m s^{-2}, and not converted. To get the *F* variable to hold the new unit, you must reassign it

```
F = F.to(u.Newton)
```

In this case, the conversion is kind of silly, because we're savvy enough to know this is actually the *definition* of the Newton, but consider

```
F = F.to(u.picoNewton)
```

Now *F* will appear in pN. It's important to notice that `astropy` is fairly smart about correctly combining units, but they're not always in the form you'd like. Consider

```
A = np.power(10.*u.centimeter, 2)
P = F/A
```

Now, *P* is actually a pressure (a force per area), but if you ask about it, you get pN cm^{-2}, and perhaps you'd prefer a pressure unit like the Pascal (Pa):

```
P.to(u.Pascal)
```

or the ever-popular pound per square inch (psi):

```
P.to(u.imperial.psi)
```

1.6 Defining Your Own Functions

Now that we've seen how to get the computer to take some of the tedium out of arithmetic and unit conversion, let's take the tedium out of computing quantities over and over, like the gravitational force. We want to think of the force as a function, where you specify the two masses and the distance between them (the *arguments* of the function), and get back the [magnitude of the] force (what the function *returns*).

`Python` lets us define functions with a notation that looks a lot like mathematical notation:

```
def grav_force(m1, m2, r):
    F = c.G * m1 * m2 / np.power(r,2)
    return F
```

Here the masses m_1, m_2 and the distance between them r are the inputs (named, naturally, `m1`, `m2`, and `r`, though you could in principle name them anything you want) and *F* is the value returned by the function, which is itself named (defined to be) `grav_force`. *G* is a constant of nature, so we don't require the user to enter it, and we make use of the fact that `astropy` has already stored it for our use. If we wanted this to be a stand-alone function we would need to ensure that `astropy` and `numpy` were imported before calculating *F* (you can, for example, import modules inside functions). As written, our function does not require that m_1, m_2, and r to have units, but if they do, F will have the correct dimensions. The equivalent mathematical notation would be

$$F(m_1, m_2, r) = G\frac{m_1 m_2}{r^2} \tag{1.3}$$

It's important to understand that, just like our mathematical function Equation 1.3, the Python function grav_force is an abstract object: it *could* be calculated for any values of its input, but as it stands it is pure possibility: no actual numbers have been computed. Likewise, the variable F that we defined inside the function grav_force is only defined while the function is being executed—if you try to print the value of F after defining and running grav_force, you'll get an error (unless you assigned the value returned by grav_force to a variable called F).

To actually use this function, we need to pick a set of values for which we want to evaluate it, and then also have a place to store the result

Let's try it!

Question 8. Compare the gravitational force of Jupiter (in Newtons) on you to the gravitational force of a member of your group. State explicitly your assumptions about the mass and distance of the objects. It is helpful to know there is a constant c.M_jup for the mass of Jupiter, and that distance to Jupiter is roughly given by 5.*u.AU, that is, about 5 astronomical units. (How far is that? Convert it to meters!)

An Introduction to Astrophysics with Python
Stars and planets
James Aguirre

Chapter 2

Kepler's Laws of Planetary Motion

We review some observational facts about our solar system and introduce Kepler's Laws of planetary motion, the first laws introduced to
describe the motions of planets. Because the orbits of planets are ellipses,
we spend some time introducing the mathematical properties of ellipses
and continue our exploration of `Python` by making plots and graphs
using `matplotlib`.

2.1 The Solar System and Early Astronomy

You are all hopefully aware that we live in a solar system of eight[1] planets orbiting
the Sun due to their mutual gravitational attraction. The inner four planets
(Mercury, Venus, Earth, and Mars) are "rocky" (composed primarily of silicon,
oxygen, and iron) and are packed quite tightly (comparatively speaking), and the
four gas (primarily hydrogen) giant planets (Jupiter, Saturn, Uranus, and Neptune)
are considerably more spaced out. Of course, in addition to these eight planets, there
are the dwarf planets,[2] asteroids (including the belt between Mars and Jupiter), a
host of objects out beyond Neptune (the imaginatively named trans-Neptunian
objects or TNOs), as well as the source of comets in the Oort Cloud extending a
good fraction of the way to the next nearest star. The modern view of our solar
system is summarized in Figures 2.1 and 2.2.

These facts about the solar system have, of course, not always been known. The
only objects visible to our ancestor's naked eyes (which, until Galileo invented the
telescope in 1609, was all we had) are the Sun, Moon, the five nearest planets

---

[1] Where's Pluto?
[2] There you are, Pluto!

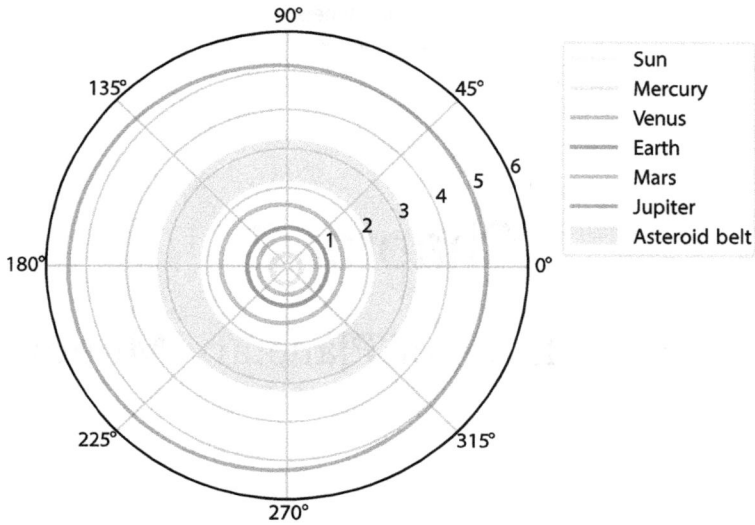

Figure 2.1. A view of the planetary orbits and rotations in inner solar system, out to the orbit of Jupiter. The actual size of the planets are far too small to show at this scale. Note the sense of the orbital rotations are all counter-clockwise when viewed from above the Earth's north pole, as in this figure, with the orbital angle increasing counter-clockwise from zero. This is also true of the planets' revolution around their own axes (except for Venus). The extent of the asteroid belt is shown, but not shown are the Trojan asteroids which lead and trail Jupiter in its orbit. The distance from the Sun (strictly, the center of mass of the solar system) is given in astronomical units (AU). The positions are computed using the `astropy.coordinates` function `solar_system_ephemeris` and the `jplephem` package to get ephemerides (positions) of solar system bodies. **This figure is also available as a movie** online at www.doi.org/10.1088/2514-3433/ade5f6.

(Mercury, Venus, Mars, Jupiter, and Saturn), the occasional comet or meteor, and the "fixed" stars. For thousands of years, it was commonly supposed that all these other objects revolved around the Earth, with the exception of comets and meteors, which were assumed to have something to do with our atmosphere (hence the same etymology for "meteor" and "meteorology"). It is easy to laugh at this view now, but the actual motions of the objects are actually quite difficult to puzzle out. There was of course the daily trip of the Sun around the Earth, which seems simple to explain as it circling us, but the attentive soon noticed that it appeared in different places relative to the fixed stars (its motion through the constellations of the Zodiac), and since the fixed stars appeared to circle us (they clearly rise and set over the course of a night), the Sun and the fixed stars can't be orbiting us at the same rate. In addition, the Moon was doing another dance on a "moon"-thly timescale, and the planets … well, they were moving around a lot relative to the fixed stars in ways that didn't make a whole lot of sense. A whole system was concocted of planets going around the Earth in circles (which were assumed to have "perfect" forms appro-priate to the heavens), but also making additional circular motions around those paths, and so on. By the middle of the sixteenth century, this whole system was quite

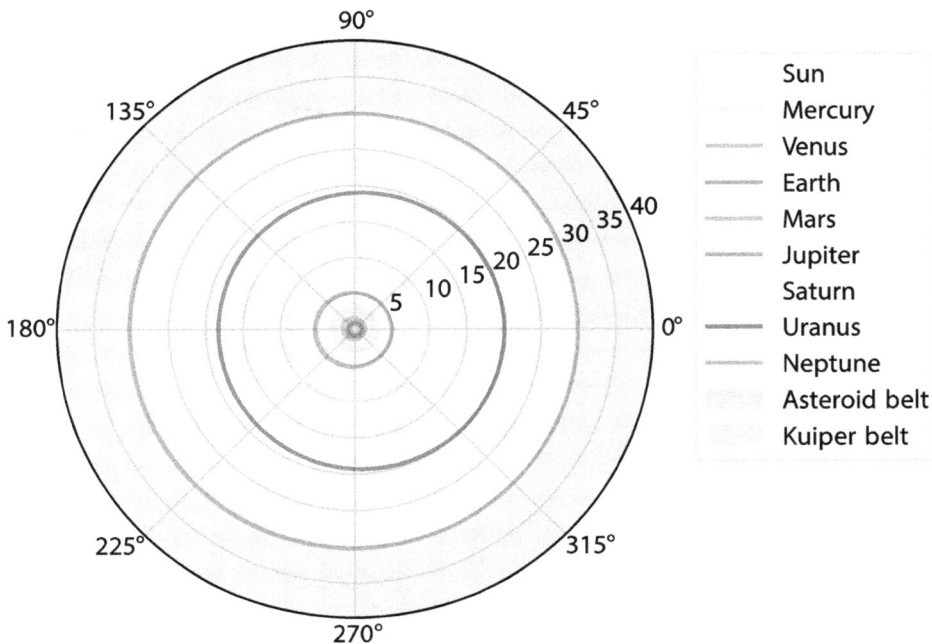

Figure 2.2. A zoom out of the solar system, showing the Kuiper Belt (containing the Trans-Neptunian objects) and the Oort Cloud. Note that the orbits of the four inner planets are all contained in a very small region near the center. The orbits of all eight of the major planets lie very nearly in the same plane and are nearly circular, so the representation is reasonable here. The Kuiper Belt is more or less in this same plane, and shares the same sense of orbital rotation. The Oort Cloud, however, has objects orbiting with a wide range of distances, eccentricities (see Equation (2.7) for the definition of "eccentricity"), and orientations relative to the plane of the planets' orbit, and is vastly larger than the scale shown here. This figure is also available as a movie online at www.doi.org/10.1088/2514-3433/ade5f6.

convoluted, if fairly accurate in predicting planetary positions. You can see an example of how difficult the problem was by looking at Figure 2.3.

Nicholas Copernicus gets most of the credit for simplifying this picture by correctly noticing that a lot of the confusion could be straightened out by putting the Sun at the center, but it should be pointed out that his system still didn't predict the positions of the planets particularly accurately, since he continued to assume they traveled around the Sun in circles. Johannes Kepler was the first to realize that the orbits of the planets around the Sun were not circles, nor circles upon circles, but actually *ellipses*. And he was able to further give rules for how the planets move around these ellipses. This was the first step on the path to actually explaining what was going on with SCIENCE™ (namely, physics).

2.2 A Bit about Ellipses

Since all of Kepler's Laws depend on properties of ellipses, it behooves us to carefully define them and their properties. It turns out that there are a number of

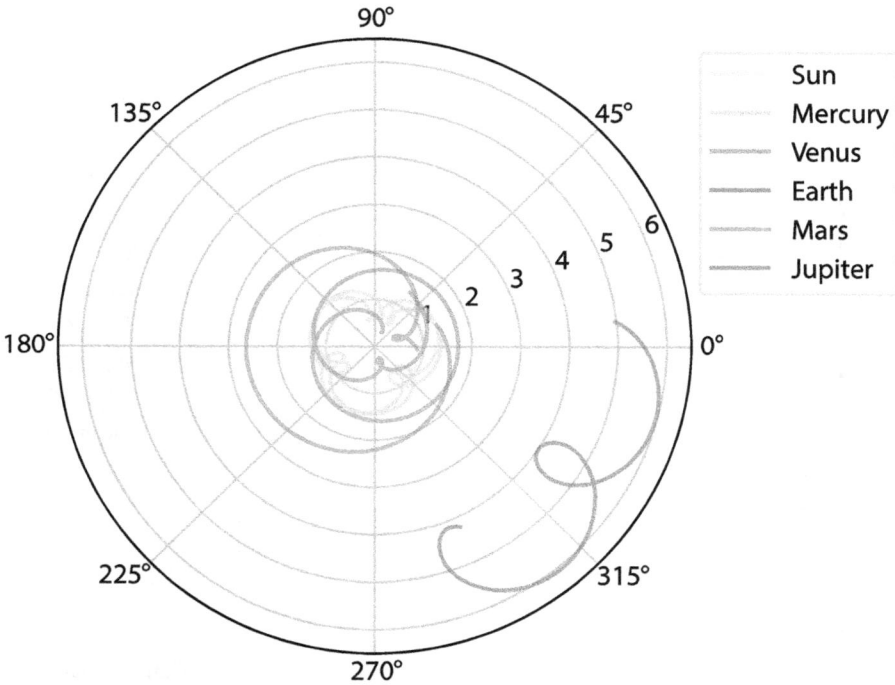

Figure 2.3. A geocentric view of the orbits of the inner planets, the Sun, and Jupiter using modern values for the orbits. Distances are shown in astronomical units (AU), and angles are as if viewing the solar system from above the Earth's north pole. In this case, we treat the Earth as if it were fixed, and show the positions of all the planets relative to it. This is how the ancients would have see the motions (except that the absolute distance scale was not known to them). The "retrograde" motion (looping back upon itself) motion of Mars and Jupiter is clearly shown. Notice that although the paths appear to cross each other, this is a reflection of the relative distance of the body to Earth at a given time, and not reflective of their path in space intersecting. Note that the asteroid belt is not shown here, as its representation in geocentric coordinates is quite complicated. This figure is also available as a movie online at www.doi.org/10.1088/2514-3433/ade5f6.

different ways to think about defining an ellipse, and different ways of defining its properties, which are handy for different purposes.

One way to define the equation for an ellipse is in Cartesian coordinates, as the set of all points (x, y) that satisfy

$$\frac{x^2}{a^2} + \frac{y^2}{b^2} = 1 \tag{2.1}$$

for two positive constants a, called the *semimajor axis* and b, the *semiminor axis*. To describe a circle requires only one number (its radius), but an ellipse requires two.[3]

---

[3] Note that these two numbers define the shape of the ellipse, but I can also shift or rotate in the plane, and this requires additional information to describe the *orientation* of the ellipse relative to the origin and the x- and y-axes.

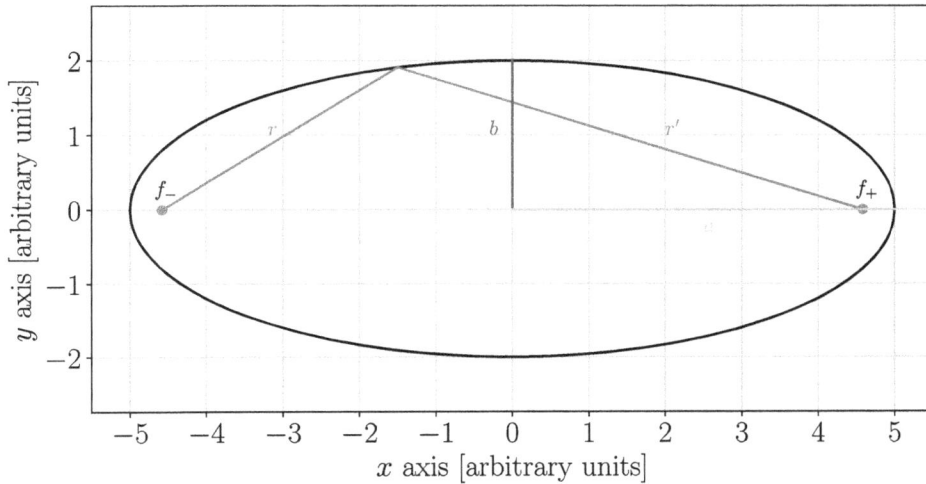

Figure 2.4. An ellipse, showing the curve itself in black, the semimajor axis a in orange, the semiminor axis b in blue, and the two foci f_+ and f_- as red dots. The sum of the distances, r and r' (green) from the foci of the ellipse to any point is equal to $2a$.

The semimajor and minor axes are one choice. Referring to Figure 2.4, we can see that the semimajor axis a is the distance between the origin and the points on the ellipse with $y = 0$. Similarly, the semiminor axis b is the distance between the origin and the points on the ellipse with $x = 0$.

We also note that there's another way to define an ellipse, which is to define two points (the *foci*) and assert that—just as a circle is the collection of points a given distance a from a point—the ellipse is the collection of points such that the sum of the distances from the two foci to any point is equal to $2a$. Again referring to Figure 2.4, this means that

$$r + r' = 2a \qquad (2.2)$$

where r is measured from one focus to *any* point on the ellipse, and r' is measured from that same point to the other focus. Figure 2.4 shows this for one particular point on the ellipse.

Now, it's not immediately obvious from looking at either Equation (2.1) or (2.2) where the foci in Figure 2.4 should be on the x-axis. However, we can notice that if the line from the focus f_+ is drawn to the point where the semiminor axis intersects the ellipse, that distance (the hypotenuse of the triangle) will be $r = a$ (since $r' + r = 2a$ and the distance from the other focus is the same by symmetry). Then, using the Pythagorean theorem, we can find the distance from the center of the ellipse to the focus as

$$f_+^2 + b^2 = a^2 \qquad (2.3)$$

which can be re-written as

$$f_+^2 = a^2 - b^2 \tag{2.4}$$

$$= a^2\left(1 - \frac{b^2}{a^2}\right) \tag{2.5}$$

from which we can write

$$f_+ = a\sqrt{1 - \frac{b^2}{a^2}} \tag{2.6}$$

To simplify this expression, we define another quantity, the *eccentricity*, related to a and b by

$$e = \sqrt{1 - \frac{b^2}{a^2}} \tag{2.7}$$

Notice that this e is *not* the Euler constant $e \approx 2.71828...$, a bit of unfortunate (if conventional) ambiguity in our notation. The context usually makes clear what is meant. Notice that for a circle (a special kind of ellipse) $a = b$ and thus $e = 0$. The most "ellipse-y" an ellipse can be is if if b gets squished all the way to zero, in which case $e = 1$.[4]

Having defined the eccentricity, the two foci for the Cartesian ellipse above are, using Equation (2.6), located at $(-ae, 0)$ and $(ae, 0)$. Notice that if we have a circle, both foci are at the origin, since $e = 0$ and thus $ae = 0$. With a bit more geometry, you can show that the distance from those coordinates to a point on the ellipse and to the other focus does indeed obey Equation (2.2).[5]

We will find that polar coordinates (r, θ) turn out to be very useful in describing orbital motions, and especially for writing out Newton's equations of motion. Recall that polar coordinates are related to Cartesian coordinates by

$$r^2 = x^2 + y^2 \tag{2.8}$$

$$\theta = \arctan(y/x) \tag{2.9}$$

where r is the distance from the origin, and θ is the angle from the positive x-axis, measured counterclockwise.

You may be less familiar (or not at all) with the equation for an ellipse in polar coordinates:

$$r(\theta) = \frac{a(1 - e^2)}{1 + e\cos(\theta)} \tag{2.10}$$

---

[4] This would actually give a straight line from $-a$ to a, which is a very special kind of ellipse.
[5] Try it!

where a is again the semimajor axis, and e is the eccentricity. Again, we need two numbers to define the shape of the ellipse: here a and e instead of a and b in Equation (2.1).

Notice that, for a circle, $a = b$, which leads to the equation in polar coordinates

$$r = a \tag{2.11}$$

(for any value of θ) and in Cartesian

$$x^2 + y^2 = a^2 = r^2 \tag{2.12}$$

which is actually the *same* equation. However, the circle is too special to see if these two forms of the ellipse equation are really the same. In fact, Equation (2.10) puts one of the foci at the origin, and the other somewhere else[6], whereas in Equation (2.1), *neither* focus is at the origin.

2.3 Kepler's Laws

Having understood some features of ellipses, we're now in a position to understand Kepler's Laws, which he formulated three laws to describe both the shape of the planets' orbits, as well as the speed at which they move around those orbits. Kepler's three laws for the motion of the planets around the Sun are:

- Planets move in elliptical orbits with the Sun at one focus of the ellipse. Nothing is at the other focus.
- The motion of the planet sweeps out equal areas of the ellipse (as measured from the focus) in equal time. Given the geometry of the ellipse, this implies that the planet moves fastest when it is closest to the Sun (perihelion) and slowest when it is furthest away (aphelion).
- The period P of a planet's orbit (the time it takes to complete one orbit) and the semimajor axis a of the orbit are related by

$$P^2 = Ka^3 \tag{2.13}$$

where K is a constant. For our solar system, this constant is

$$K = \frac{(1\,\text{year})^2}{(1\,\text{AU})^3} \tag{2.14}$$

where $1\,\text{AU} = 150 \times 10^6$ km.

Kepler's Laws are wholly *empirical* (Kepler figured them out based on observations, and he didn't say anything about *why* these facts are true), and he also doesn't give us (from the laws themselves) some of the numbers we are interested in. For example, the laws don't tell us the size a or the orientation in space of the elliptical orbits of the

---

[6] Can you figure out where?

planets; we have to figure that out from observations. (Kepler actually spent a lot of time trying to figure out why the planets were at the distances from the Sun that they are, without any success.) And the theory doesn't tell us the value of K; that too has to come from observations. What can be measured less directly by patient observers is the period P, the time it takes the planets to go around the Sun once.

One of the goals of this first part of the book is to try to show how a very nice summary of experimental results, like Kepler's Laws, can actually be *explained* in a "deeper" way by physical theory. "Deep" is a loaded word, but here I mean that the explanation of the laws will arise from a theory that is more general (it can handle other kinds of behavior than the motions of planets in our solar system) and it relates its constants back to some fundamental physical parameters of the system (like the masses of the planets and the Sun).

Our next exercise will spend time exploring the two forms of the ellipse equations, Equations (2.1) and (2.10).

2.4 Exercise: Kepler's Laws and Ellipses

We'll now work through some exercises to
- Become comfortable with the properties of ellipses in the context of Kepler's Laws of Planetary Motion
- Become familiar with `Python` programs for plotting, in both Cartesian and polar coordinate systems
- Improve your ability to write functions in `Python`

Recall the definitions of the equations of ellipse in Cartesian (Equation (2.1)) and polar (Equation (2.10)) coordinates. While you may be more familiar with the Cartesian form of the equation of an ellipse, we'll find that it's easier to start plotting the polar form. $r(\theta)$ is easy to compute for any θ in the range of interest because we know the ellipse is a closed curve, and it must close after the polar angle has wrapped from 0 to 2π. So we readily know what range of angles to pick.

Question 1. Write a `Python` function that accepts a, e, and θ as arguments and calculates r using the polar form of the equation of an ellipse. Call this function something descriptive, like `ellipse_polar`. (Recall `grav_force` from the first in-class exercise as an example of a simple `Python` function).

Question 2. Plot `theta` on the x-axis and `r` on the y-axis using `plt.plot(theta, r)` for

$$a = 1 \tag{2.15}$$

$$e = 0.3 \tag{2.16}$$

Recall `np.linspace` for getting values of θ. Label your axes!

Explain why this plot does not look like an ellipse.

From the plot (or from your calculated values), how far away from the origin is the point of the ellipse that comes closest to the origin? What is the maximum distance away from the origin of a point on the ellipse?

Question 3. We can plot directly in polar coordinates using `plt.polar(theta,r)` (note the order of the arguments). Plot the ellipse in polar coordinates for

$$a = 1 \tag{2.17}$$

$$e = 0.3 \tag{2.18}$$

and $0 \leqslant \theta < 2\pi$. Again, by reference to the plot, at what θ is the ellipse closest to the origin? At what θ is it furthest away? (It may help to put a point at the origin with `plt.plot(0,0,'ro')`.)

Your function for $r(\theta)$ was pretty simple: values for θ (and a and e) in, and r out. So the `return` statement in the function was also simple. If we want to write a function to calculate the equation of an ellipse in Cartesian coordinates, we could try re-writing

$$y = \pm b \sqrt{1 - \frac{x^2}{a^2}} \tag{2.19}$$

but note that $y(x)$ is not single-valued for any given value of x, so we'd have to decide how to return a y value given an x. Since we want the values for plotting purposes, let's try a different approach: generating matched (x, y) lists for some given number of points in the plot. The first question we want to address is below.

Question 4. Using the Cartesian form of the ellipse, what are the maximum and minimum values of x which lie on the curve (in terms of a and b)? Show your reasoning. This part is *not* numerical.

Here I've written an example that doesn't allow the user to input x, but just a, b and the number of points desired, and out come two lists, both stored in a single variable:

```
def ellipse_cartesian_dict(a, b, num=100):
    """ Plot an ellipse in Cartesian coordinates """

    # Evaluate the positive valued solution
    x = np.linspace(-a,a,num=num)
    y = b*np.sqrt(1-np.power(x/a,2))
    #Join together the positive and negative solutions for
    y, reversing the order for the negative solutions
    # This is to make the plotting nice, by following the
    curve of the ellipse around continuously
    y_all = np.concatenate((y,np.flip(-y,axis=0)))
    # Now make an appropriate x array for the y values
    x_all = np.concatenate((x,np.flip(x,axis=0)))

    return {'x':x_all,'y':y_all}
```

Let's look at a couple of features here. One is that you can enter the variable num (the number of points for which the equation is evaluated) by typing

```
ellipse = ellipse_cartesian_dict(5, 2, num=50)
print(ellipse['x'].shape)

(100,)
```

Notice that I asked for 50 points, but it actually gave me twice as many (50 for positive y and 50 for negative y). If I had entered nothing for num (the "default"), I would have gotten

```
ellipse = ellipse_cartesian_dict(5, 2)
print(ellipse['x'].shape)

(200,)
```

Looking at the function in greater detail, the first two lines should make sense, and while the next two look cryptic, they are just producing numpy arrays of x, y for both positive and negative solutions of y, ordered so they will plot nicely. A final thing to note is that the object that is returned is a *dictionary*, and you can access the item named x by typing ellipse['x']. Dictionaries in Python are very powerful, and many functions make use of them. The other way we could have done this is by writing the same function with a slightly different return statement:

```
def ellipse_cartesian_tuple(a, b, num=100):
    """ Plot an ellipse in Cartesian coordinates """

    # Evaluate the positive valued solution
```

where (x,y) is a *tuple*, and packages the two variables together in a new object. Note that with the tuple, there is no explicit label for the variables (you just have to "know" which one is which), so in that sense, the dictionary is more comprehensible to humans (but requires more typing). However, for users of MATLAB, the tuple allows a familiar construction:

```
    x = np.linspace(-a,a,num=num)
    y = b*np.sqrt(1-np.power(x/a,2))
    #Join together the positive and negative solutions for
    y, reversing the order for the negative solutions
    # This is to make the plotting nice, by following the
    curve of the ellipse around continuously
    y_all = np.concatenate((y,np.flip(-y,axis=0)))
    # Now make an appropriate x array for the y values
    x_all = np.concatenate((x,np.flip(x,axis=0)))

    return (x_all, y_all)
```

Which form you use is up to you.

One last thing to note about the function: the comment in triple quotes is quite useful. Notice what happens when we type:

```
ellipse_cartesian_dict?
```

Question 5. Plot the same ellipse as in Question 2 ($a = 1, e = 0.3$) using the Cartesian form. First write a function to compute b from a and e, so the two ellipses really are the same. Plot a point marking the origin, and label your axes.

An Introduction to Astrophysics with Python
Stars and planets
James Aguirre

Chapter 3

Polar Coordinates and Kepler's Second Law

We build on the notion of polar coordinates introduced for specifying the shape of an ellipse to using them for describing motion in the plane, and in particular the position and velocity of a planet. We then discuss Kepler's Second Law, and by introducing the notion of numerical evaluation of a definite integral, we use this to explore the area of ellipses, both numerically and graphically, which we relate to Kepler's Second Law. We also continue to gain practice with making and interpreting plots.

We've given equations for ellipses in both polar (r, θ) and Cartesian (x, y) coordinates, namely

$$r(\theta) = \frac{a(1 - e^2)}{1 + e \cos(\theta)} \tag{3.1}$$

and

$$\frac{x^2}{a^2} + \frac{y^2}{b^2} = 1 \tag{3.2}$$

which describe the same shape if we set

$$b = a\sqrt{1 - e^2} \tag{3.3}$$

Let's recall the relation between the two systems, first going from Cartesian to polar:

$$r = \sqrt{x^2 + y^2} \tag{3.4}$$

$$\theta = \arctan y/x \tag{3.5}$$

doi:10.1088/2514-3433/ade5f6ch3

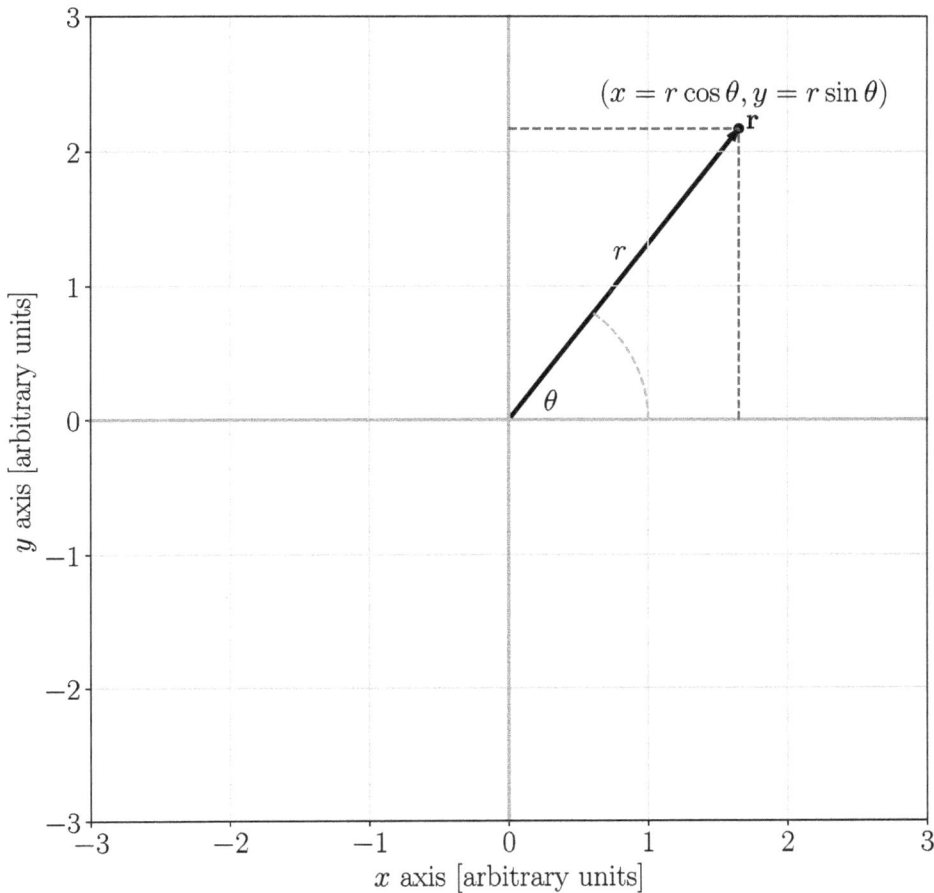

Figure 3.1. The definition of polar coordinates, showing the position vector $\mathbf{r} = r\hat{\mathbf{r}} = x\hat{\mathbf{x}} + y\hat{\mathbf{y}}$.

and back again

$$x = r\cos(\theta) \tag{3.6}$$

$$y = r\sin(\theta) \tag{3.7}$$

Notice that these relations follow from geometry, and are illustrated in Figure 3.1.

3.1 Kinematics in Polar Coordinates

Particle motion in Cartesian coordinates is straightforward to describe, and is generally familiar from introductory physics. However, we will find that all the theoretical results for the problem of two bodies interacting via gravity will be more conveniently couched in terms of polar coordinates, so we now want to think of how to describe the kinematics[1] in polar coordinates. In Cartesian coordinates, we can describe the path of a particle in the plane with the vector

$$\mathbf{r} = x(t)\hat{\mathbf{x}} + y(t)\hat{\mathbf{y}} \tag{3.8}$$

Now, the Cartesian unit vectors have the nice property that no matter where we are in the (x, y)-plane, they always point in the same direction, namely, along the x- or y-axes; in other words, they do *not* depend on the functions $x(t)$, $y(t)$, or t itself.

Let's see what happens if we use this fact to directly compute the time derivative[2] of $\mathbf{r}$ using the chain rule for products:

$$\frac{d\mathbf{r}}{dt} = x(t)\frac{d\hat{\mathbf{x}}}{dt} + \frac{dx}{dt}\hat{\mathbf{x}} + y(t)\frac{d\hat{\mathbf{y}}}{dt} + \frac{dy}{dt}\hat{\mathbf{y}} \tag{3.9}$$

$$= \dot{\mathbf{r}} = x\dot{\hat{\mathbf{x}}} + \dot{x}\hat{\mathbf{x}} + y\dot{\hat{\mathbf{y}}} + \dot{y}\hat{\mathbf{y}} \tag{3.10}$$

$$= \dot{x}\hat{\mathbf{x}} + \dot{y}\hat{\mathbf{y}} \tag{3.11}$$

since $\dot{\hat{\mathbf{x}}} = \dot{\hat{\mathbf{y}}} = 0$. This is so trivial that it may never have even occurred to you that you should take the derivative of a unit vector when calculating the time derivative of a vector.

However, in polar coordinates it turns out that this nice property that the time derivatives of the unit vectors are zero *doesn't* hold: $\hat{\mathbf{r}}$ and $\hat{\boldsymbol{\theta}}$ point in different directions depending on where we are in the (x, y)-plane, and so their time derivatives *won't* be zero. Referring to Figure 3.2, the point in question has associated with it the unit vectors $\hat{\mathbf{r}}$ and $\hat{\boldsymbol{\theta}}$, which are different at different points in the plane. To see this explicitly, let's use the relation between Cartesian and polar coordinates, but include all of the time dependence:

$$x(t) = r(t)\cos\theta(t) \tag{3.12}$$

$$y(t) = r(t)\sin\theta(t) \tag{3.13}$$

so we can write

$$\mathbf{r}(t) = x(t)\hat{\mathbf{x}} + y(t)\hat{\mathbf{y}} \tag{3.14}$$

$$= r(t)\cos\theta(t)\hat{\mathbf{x}} + r(t)\sin\theta(t)\hat{\mathbf{y}} \tag{3.15}$$

$$= r(t)(\cos\theta(t)\hat{\mathbf{x}} + \sin\theta(t)\hat{\mathbf{y}}) \tag{3.16}$$

$$\equiv r(t)\hat{\mathbf{r}} \tag{3.17}$$

where the last line *defines* the unit vector $\hat{\mathbf{r}}$. Notice that indeed

$$\hat{\mathbf{r}} = \cos\theta\hat{\mathbf{x}} + \sin\theta\hat{\mathbf{y}} \tag{3.18}$$

---

[1] *Kinematics* here is a fancy word to mean a quantitative description of the position, velocity, acceleration, etc. of particles, without regard to the physical laws that determine those quantities.
[2] The notation here of one dot over the variable means one time derivative: $\dot{x} = \frac{dx}{dt}$.

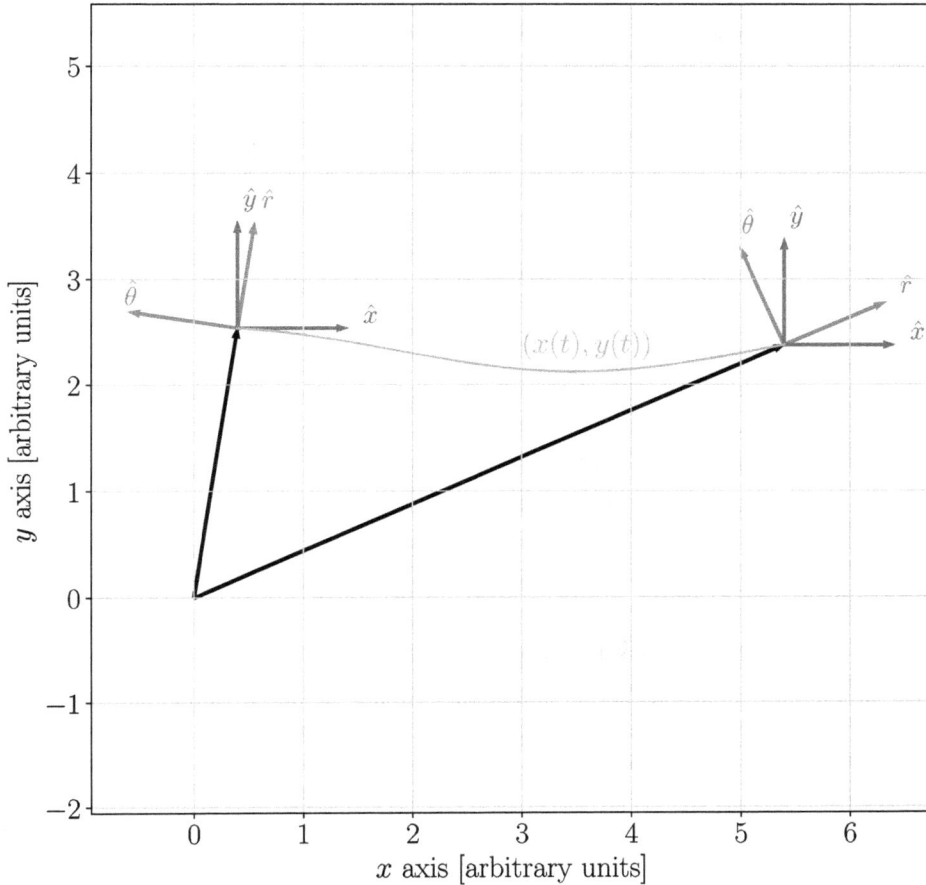

Figure 3.2. An illustration of how the polar coordinate unit vectors change with time in the description of the motion of a particle. We imagine a particle moving along the trajectory $(x(t), y(t))$ shown in gray. Notice that time is not explicitly shown in this plot (the axes are the coordinates x and y), but implicitly is used to describe where along the trajectory $(x(t), y(t))$ the particle is. The position vectors $\mathbf{r}$ are show at two times along the path, along with the Cartesian (blue) and polar (orange) unit vectors used to decompose the position vector. Note that the Cartesian unit vectors always maintain the same orientation relative to the coordinate axes, whereas the polar unit vectors reflect the radial and tangential directions *at the specific point*.

is a unit vector (because $\cos^2 \theta + \sin^2 \theta = 1$) that always points radially outward in a direction specified by $\theta(t)$. The dependence of $\hat{\mathbf{r}}$ on θ, which then depends on t, is why when we take its time derivative we won't get zero! The unit vector orthogonal to $\hat{\mathbf{r}}$ is $\hat{\boldsymbol{\theta}}$, which is defined by

$$\hat{\boldsymbol{\theta}} = -\sin\theta\hat{\mathbf{x}} + \cos\theta\hat{\mathbf{y}} \qquad (3.19)$$

These have the desired properties[3] that $\hat{\mathbf{r}} \cdot \hat{\boldsymbol{\theta}} = 0$ and $\hat{\mathbf{r}} \cdot \hat{\mathbf{r}} = \hat{\boldsymbol{\theta}} \cdot \hat{\boldsymbol{\theta}} = 1$.

---

[3] You should verify that you understand why these are true!

For the time derivative of **r** expressed in terms of $\hat{\mathbf{r}}$ and $\hat{\boldsymbol{\theta}}$, let's write explicitly

$$\dot{\mathbf{r}} = \dot{x}\hat{\mathbf{x}} + \dot{y}\hat{\mathbf{y}} \tag{3.20}$$

$$= (-r\sin\theta\dot{\theta} + \dot{r}\cos\theta)\hat{\mathbf{x}} + (r\cos\theta\dot{\theta} + \dot{r}\sin\theta)\hat{\mathbf{y}} \tag{3.21}$$

$$= \dot{r}(\cos\theta\hat{\mathbf{x}} + \sin\theta\hat{\mathbf{y}}) + r\dot{\theta}(-\sin\theta\hat{\mathbf{x}} + \cos\theta\hat{\mathbf{y}}) \tag{3.22}$$

$$= \dot{r}\hat{\mathbf{r}} + r\dot{\theta}\hat{\boldsymbol{\theta}} \tag{3.23}$$

$$\equiv v_r\hat{\mathbf{r}} + v_\theta\hat{\boldsymbol{\theta}} \tag{3.24}$$

where the last line *defines* the radial velocity $v_r = \dot{r}$ and the tangential velocity $v_\theta = r\dot{\theta}$.

We can also compute the squared magnitude of the velocity as

$$|\dot{\mathbf{r}}|^2 = \mathbf{r}\cdot\mathbf{r} = \dot{r}^2 + r^2\dot{\theta}^2 \tag{3.25}$$

We could also apply similar reasoning to obtain the acceleration, but will leave this exercise to the reader.

3.2 An Application to Kepler's Second Law

So far we have primarily focused on Kepler's First Law (that orbits are ellipses) and getting comfortable with the properties of the ellipse (and, of course, how to make plots and do calculations with Python). But now let's look at the Second Law, that the motion of the planet sweeps out "equal areas in equal times." To make use of this, we're going to need to figure out how to calculate the area of an ellipse. Calculating an area means we need to do an integral, so we need to find a differential bit of area of the ellipse. We can start by calculating the area of a triangle from one focus to a point on the ellipse and another point after rotating by $\Delta\theta$, where the sides are r and (approximately) $r\Delta\theta$ as shown in Figure 3.3.

Note that if we think of the area as a triangle, the side is length $r\Delta\theta$ using the small angle approximation of the exact value $r\tan\theta$, and if we think of measuring around a circle, the arc length is exactly $r\Delta\theta$. In either case, this little bit of area is

$$\Delta A \approx \frac{1}{2}r(r\Delta\theta) = \frac{1}{2}r^2\Delta\theta \tag{3.26}$$

This is of course wrong for any finite value of $\Delta\theta$, but we can take the differential limit

$$\frac{dA}{d\theta} = \lim_{\Delta\theta\to 0}\frac{\Delta A}{\Delta\theta} = \frac{\frac{1}{2}r^2\Delta\theta}{\Delta\theta} = \frac{1}{2}r^2 \tag{3.27}$$

We can check that this reasoning clearly gives the right answer for a circle with $r = a$ as

$$A_{\text{circle}} = \int dA_{\text{circle}} = \int_0^{2\pi}\frac{1}{2}r^2 d\theta = \frac{1}{2}a^2\int_0^{2\pi}d\theta = \frac{1}{2}a^2(2\pi) = \pi a^2 \tag{3.28}$$

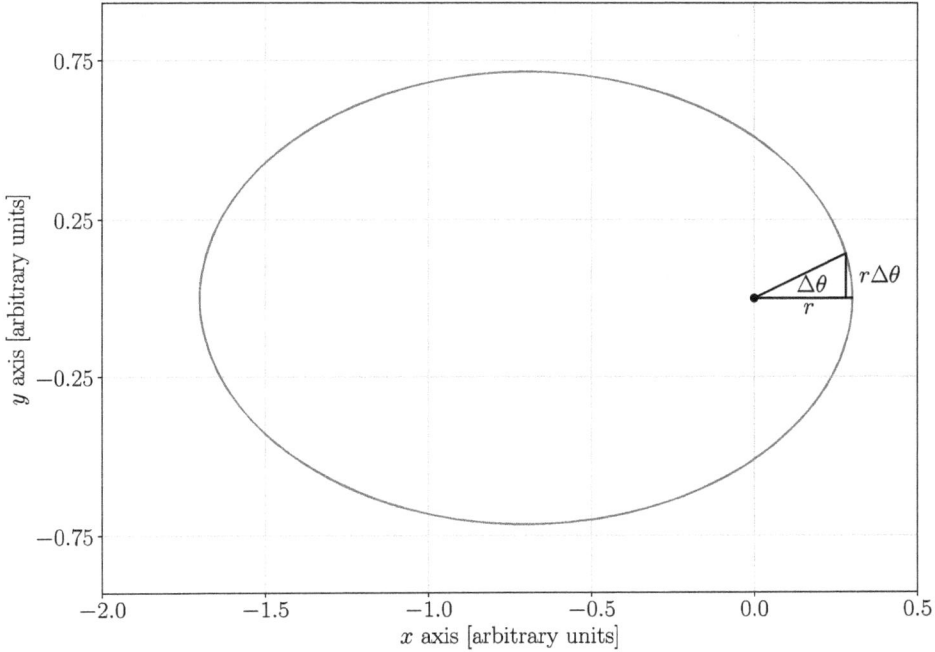

Figure 3.3. Calculating the area of an ellipse by defining an element of differential area in polar coordinates.

If we put in the explicit equation for an ellipse, we would get the scary integral

$$A_{\text{ellipse}} = \int_0^{2\pi} dA_{\text{ellipse}} = \int_0^{2\pi} \frac{1}{2} r^2(\theta) d\theta = \int_0^{2\pi} \frac{1}{2} \left(\frac{a(1 - e^2)}{1 + e \cos(\theta)} \right)^2 d\theta \qquad (3.29)$$

It turns out that the value of this integral has a simple formula,[4] namely

$$A_{\text{ellipse}} = \pi a b \qquad (3.30)$$

which reduces to the right form for $e = 0 \Rightarrow a = b$, of course.

The above statements are just about geometry, but we can relate this reasoning back to the motion of the particle by using Kepler's Second Law, which we can think of as saying

$$\frac{dA}{dt} = \text{constant} \qquad (3.31)$$

We can relate $\frac{dA}{d\theta}$ to $\frac{dA}{dt}$ by the chain rule

$$\frac{dA}{dt} = \frac{dA}{d\theta} \frac{d\theta}{dt} = \frac{1}{2} r^2 \frac{d\theta}{dt} = \text{constant} \qquad (3.32)$$

---

[4] Which is much easier to evaluate in Cartesian coordinates, as it turns out.

which, we're going to find, is basically the statement of the conservation of angular momentum, since this quantity is proportional to the angular momentum.

3.3 Calculating Integrals Numerically

We'd now like to see how to evaluate the integral Equation (3.29) *numerically* (rather than trying to be clever to get the analytic form). In this case, someone has figured out how to the integral, so we have an "analytic form" to check against, and don't need to do the integral numerically, but having this check will allow us to test our numerical integration. In general, integrals *can't* be done analytically, so having the ability to evaluate numerical approximations is a useful tool to have in our belt.

The Scientific Python (scipy) module implements a version of the Riemann sum[5] known as the "trapezoidal rule" in

```
scipy.integrate.trapezoid(f, x=x)
```

where x is a list of numbers representing the independent variable and f is a list of the corresponding values of the function to integrate. The limits of the integral are *implicit*, that is, they're encoded in the first and last points you chose for x (and f). Thus, you want to give trapezoid arrays corresponding only to the range of points you're actually interested in integrating. scipy also implements a version of the indefinite integral as as

```
scipy.integrate.cumulative_trapezoid(f, x=x, initial=0)
```

which returns a function effectively corresponding to the anti-derivative. Let's look at how the syntax of the Python command

```
F = scipy.integrate.cumulative_trapezoid(f, x=x, initial=0)
```

translates to the mathematical version

$$F(t) - F(t_i) = \int_{t_i}^{t} f(s)ds \tag{3.33}$$

(note the use of the dummy integration variable s).

First of all, we'd like the array F to be the same length as f, but because the trapezoidal rule averages successive points, a list of N function values will only produce $N - 1$ integrated values. The statement initial = 0 says to make the first value of F=0, which is usually what you want, since mathematically

$$\int_{t_i}^{t_i} f(s)ds = 0 \tag{3.34}$$

Then we can make the following identifications:

$$F(t) = \text{F} \tag{3.35}$$

---

[5] Recall this from calculus? We will review it again shortly.

$$f(t) = \texttt{f} \tag{3.36}$$

$$t_i = \texttt{t[0]} \tag{3.37}$$

$$t = \texttt{t[-1]} \tag{3.38}$$

Note the `Python` way to indicate the last element in the list. Also notice that dt doesn't appear explicitly in the call to `cumulative_trapezoid`; it is computed "under the hood" from `t`. That accounts for everything, except for $F(t_i)$. In fact, we've assumed it's zero, and we should really write `F = scipy.integrate.cumulative_trapezoid(f, x=x, initial=Fi)` to properly incorporate $F(t_i)=$`Fi`. This is very important, since it's how we get the initial condition included! (Notice that putting `F-Fi` on the left hand side of the `Python` statement would look more like the calculus version, but is nonsense code: I want to assign everything on the right to the variable name on the left.)

To get both `trapezoid` and `cumulative_trapezoid` we needed to import the `scipy.integrate` module; here we will show how to import just these two methods from that module and give them the names `trapezoid` and `cumulative_trapezoid`:

```
from scipy.integrate import trapezoid, cumulative_trapezoid
```

3.4 Summary

The key ideas from this reading are:
- Polar coordinates are going to be very useful in the study of the interaction of two bodies under gravity, but we have to take some care when defining quantities like velocity and acceleration in these coordinates.
- We can view Kepler's Second Law quantitatively as a statement about the conservation of a quantity which will turn out to be angular momentum.
- There are numerical methods for integrating functions which are expressed as numerical values corresponding to a set of independent variable values, without having to specify the analytic form of the function.

3.5 Exercise: Area and Other Properties of Ellipses

For this exercise, our goals are to
- Continue gaining confidence with using functions in `Python` and using the results of calculations to make plots
- Learn about a `Python`-based method for computing a difficult integral numerically
- Apply these techniques to explore the meaning of Kepler's Second Law of planetary motion
- Become more familiar with the expression of functions in both polar and Cartesian coordinates

For this exercise, we're going to re-use the functions you wrote in Exercise 2.4. Use the values

$$a = 1 \qquad (3.39)$$

$$e = 0.7 \qquad (3.40)$$

$$b = a\sqrt{1 - e^2} \qquad (3.41)$$

Question 1. Calculate the function

$$\Delta A(\theta) = \frac{1}{2}r^2(\theta)\Delta\theta \qquad (3.42)$$

for an ellipse using `ellipse_polar` (or your version of this function) and the values of a and e above for a range of θ, and plot ΔA versus θ. (Label axes.) Since we are going to use this function for numerical integration, we want to find a subdivision of θ, so use `num=10000` in `np.linspace`. You are plotting how much new area is added to the ellipse for each equal step of $\Delta\theta$. Notice that I wrote ΔA instead of dA because you are calculating this quantity for a finite step $\Delta\theta$.

Question 2. Use Kepler's Second Law to answer these questions. When ΔA is at its maximum, is the time interval to travel a step of $\Delta\theta$ larger or smaller than when ΔA is at its minimum? Explain why. Where, in your plot in Question 1, would the planet on this elliptical orbit be traveling the *slowest*? Explain why. No calculations are required here, but explain your answer in words (and/or equations).

Keep in mind: in your plot, each new point of ΔA doesn't correspond to an equal step in *time* for a planet following this orbit, but an equal step in θ (polar angle).

Question 3. Plot the cumulative area $A(\theta)$ of the ellipse as a function of θ using `cumulatize_trapezoid` from `scipy.integrate`. Explain why the shape makes sense given your plot of ΔA above. Is the final value of A equal to the expected area? Note that ordinarily we would integrate a function $f(\theta)$ using `cumulative_trapezoid(f, x=theta)` in order to approximate

$$\int_0^{2\pi} f(\theta)d\theta \qquad (3.43)$$

but above you've already included the $\Delta\theta$ in the integrate ΔA, so we just need `cumulative_trapezoid(dA)`.

Question 4. Write a `Python` function to accept r,θ values and return the corresponding x,y values. You can use either the tuple or dictionary form to return your values. It's always good practice to check code with cases where the answer is known. Calculate for yourself what you expect for $(r = 1, \theta = 0)$ and $(r = 3, \theta = \pi)$, and verify that your function works by showing the output.

Question 5. Use all the functions you've written so far to plot the polar version of the ellipse (now with r and θ converted to x and y) and the Cartesian version on the same (Cartesian) axes. You can use `plt.grid()` to put a grid in the plot and make it clearer where the origin is, and where the extents of the two ellipses are. `plt.axis ('equal')` will also ensure that the plot makes equal lengths measured along x and y actually appear the same size on both axes. Label your axes! Explain why, even though a and e (and b) are identical for the two cases, the two ellipses do not line up on top of each other.

Question 6. What are the (x, y) positions of the two foci of the ellipse for both the polar form and the Cartesian form? (It's a bit more difficult to figure this out for the polar form than for the Cartesian.) You may find it helpful to refer to your plot in Question 5 as a guide, but you should give an analytic answer. Hint: how far, measured along the x-axis, is the focus from a point on the ellipse?

Question 7. Explain what you need to do to adjust the coordinates of either the Cartesian or polar form so they can be plotted on top of each other. Repeat the plot of Question 2 with this adjustment and show that the ellipses do line up on top of each other. You might want to use `plt.plot (x,y,'r--')` to get a dashed line for one of the forms. Mark the location of both foci on the plot using `plt.plot(x_f1, y_f1,'ro')` which plots a red dot (`'ro'`) at the location x_f1, y_f1 of one of the foci.

AAS | IOP Astronomy

An Introduction to Astrophysics with Python
Stars and planets
James Aguirre

Chapter 4

Gravity, Ordinary Differential Equations, and Non-dimensionalizing

Several important new ideas are introduced here. We continue with the notion of numerical integration but now show how it can be applied to solving the differential equations that come from formulating motion under gravity according to Newton's Laws, and particularly the problem of free fall in a gravitational field. This is the first step on the way to formulating planetary motion in terms of Newton's Laws and in explaining the underlying reasons for Kepler's Laws. We introduce some additional numerical functions in `Python`. We also introduce the idea of dimensional analysis to extract key physical insights with a minimum of effort.

4.1 Motion in a Constant Gravitational Field

Let's start with a simple problem you've hopefully seen solved before, even if not in quite the way we'll do it here. We consider an object of mass m falling under the influence of gravity, where the gravitational force will be taken to be constant and pointing in the negative x-direction,[1] that is

$$\mathbf{F} = m\mathbf{a} = m\frac{d^2\mathbf{x}}{dt^2} = -mg\hat{\mathbf{x}} \qquad (4.1)$$

This leads to three equations of motion, corresponding to each component of the vector in Equation (4.1):

---

[1] Often we denote the gravitational direction as z, i.e., vertical, but the physical content doesn't depend on how we label our axes.

doi:10.1088/2514-3433/ade5f6ch4

$$m\frac{d^2x}{dt^2} = -mg \tag{4.2}$$

$$m\frac{d^2y}{dt^2} = 0 \tag{4.3}$$

$$m\frac{d^2z}{dt^2} = 0 \tag{4.4}$$

We need to solve these equations to find the position of the particle as a function of time in three dimensions, that is, to find $x(t)$, $y(t)$ and $z(t)$. To do this, we also need *initial conditions*. These are the initial position and velocity of the particle. We'll assume the mass m is initially at rest, and is a distance x_0 away from the origin along the x-axis:

$$x(0) = x_0 \qquad \dot{x}(0) = 0 \tag{4.5}$$

$$y(0) = 0 \qquad \dot{y}(0) = 0 \tag{4.6}$$

$$z(0) = 0 \qquad \dot{z}(0) = 0 \tag{4.7}$$

Notice that our choice to point the gravitational force in the negative x-direction and place the particle at a positive x position means that we'll expect the particle to move to the left, toward the origin. Let's see how that works out.

Starting with an easy equation, we're going to integrate the equation for the y-coordinate from time 0 to time t, on both sides of the equation. Integrating the left-hand side gives

$$\int_0^t \frac{d^2y}{dt'^2}dt' = \frac{dy}{dt}(t) - \frac{dy}{dt}(0) = \frac{dy}{dt} - \dot{y}(0) = 0 \tag{4.8}$$

Notice that this uses the fundamental theorem of calculus, and the initial condition that $\dot{y}(0) = 0$. Integrating the right-hand side is even easier:

$$\int_0^t 0dt' = 0 \tag{4.9}$$

so we have

$$\frac{dy}{dt} = 0 \tag{4.10}$$

We still don't know $y(t)$, but we can integrate again

$$\int_0^t \frac{dy}{dt'}dt' = y(t) - y(0) = y(t) = \int_0^t 0dt' = 0 \tag{4.11}$$

where again we used an initial condition, this time that $y(0) = 0$. This leads to the conclusion that $y(t) = 0$ for all time. You should convince yourself that this does satisfy the differential equation and the initial conditions. A similar set of calculations would also tell us that $z(t) = 0$.

Now, let's try something harder. Looking at the x component, we have

$$\frac{d^2x}{dt^2} = -g \tag{4.12}$$

Integrating both sides as before,

$$\int_0^t \frac{d^2x}{dt'^2}dt' = \frac{dx}{dt}(t) - \frac{dx}{dt}(0) = \frac{dx}{dt} - \dot{x}(0) \tag{4.13}$$

$$= \frac{dx}{dt} = \int_0^t -g\,dt' = -gt \tag{4.14}$$

by using one of our initial conditions. Then, we integrate again to get

$$\int_0^t \frac{dx}{dt'}dt' = x(t) - x(0) = x(t) - x_0 = \int_0^t -gt'dt' = -\frac{1}{2}gt^2 \tag{4.15}$$

which leads to the expected result that

$$x(t) = x_0 - \frac{1}{2}gt^2 \tag{4.16}$$

Notice that, as you might have expected, at $t = 0$, the object is at x_0 and for all $t > 0$, the x position *decreases*.

An interesting question we'll return to is how long it takes for the object to fall from x_0 to the origin (the "free-fall time"), and the answer is simply found by noting that at the time t_{ff} when the object arrives at the origin, we must have

$$x(t_{\text{ff}}) = 0 = x_0 - \frac{1}{2}gt_{\text{ff}}^2 \tag{4.17}$$

or

$$t_{\text{ff}} = \sqrt{\frac{2x_0}{g}} \tag{4.18}$$

The appearance of a characteristic time for the problem, just depends on the constants and initial conditions, is going to recur in Section 4.3. Another way to frame the problem, which at first doesn't seem very useful, is to write it as two-first-order differential equations instead of one second-order equation. It turns out we can always do this, and it makes the problem much easier when we ask computers to help, since computers are really only good at integrating first-order equations. If we define a new variable v (suggestively named) as

$$\frac{dx}{dt} = v \qquad (4.19)$$

then Equation (4.1) becomes two equations

$$\frac{dx}{dt} = v \qquad (4.20)$$

$$\frac{dv}{dt} = -g \qquad (4.21)$$

with the associated initial conditions $v(0) = 0$, $x(0) = x_0$. This would give the same result as before, by integrating each equation and applying the initial conditions:

$$x(t) = vt + x_0 \qquad (4.22)$$

$$v(t) = -gt \qquad (4.23)$$

4.2 Numerical Solution of a Differential Equation

4.2.1 Discretizing

That was fairly straightforward, and the result probably did not come as a surprise. But how would we do the integral in Equation (4.13) numerically? I suppose I have a discrete list of values for the time

$$t = [t_1, t_2, t_3, \ldots, t_N]$$

where $a = t_1 < t_2 < t_3 \cdots < t_N = b$. Such a list for a function $f(t)$ is referred to as a *partition* of interval $[a, b]$. Concretely, let's think of these times as times at which I will evaluate the force on the right-hand side of Equation (4.12) (it's constant, so that's actually easy).

To do the first integral to get velocity, let's appeal to the basic definition of the integral, the so-called *Riemann integral*. This defines the definite integral of the function $f(t)$ to be

$$\int_a^b f(t)dt = \lim_{N \to \infty} \sum_{i=1}^{N} f(t_i)(t_{i+1} - t_i)$$

Mathematically, I assume that I can keep taking finer and finer meshes (with the t_i more closely spaced), and if this limit exists, then this is the integral. Numerically, of course, I can't evaluate a real infinite sum, and so I just make $t_{i+1} - t_i$ small enough that the difference between the value I calculate and the true value of the integral is small enough for the purposes of the calculation in question. "Small enough" in numerical calculations usually means that the error from the numerical approximations is smaller than other sources of uncertainty, such as the accuracy with our measurements are made, or the accuracy of the approximations we've made in deriving the equations to be solved numerically in the first place.

Note that the above approach will produce a value for any definite integral (assuming all of the function evaluations $f(t_i)$ are finite), but it doesn't produce a new *function*, i.e., the anti-derivative. For that, we notice that, by definition

$$\int_a^b f(t)dt = F(b) - F(a)$$

where

$$\frac{dF}{dt} = f \qquad (4.24)$$

Remember that the anti-derivative is only defined up to a constant, that is, I can replace any $F(t)$ that satisfies Equation (4.24) to $F(t) + C$ and Equation (4.24) will still be satisfied.

To find the function $F(t)$, we can evaluate

$$F(t_1) - F(a) \approx \sum_{i=1}^{1} f(t_i)(t_{i+1} - t_i)$$

and

$$F(t_2) - F(a) \approx \sum_{i=1}^{2} f(t_i)(t_{i+1} - t_i)$$

and so on to generate $F(t)$ for every t_i in my original list. Note that this is equivalent to choosing the integration constant C such that $C = F(a)$.

4.2.2 Numerical Solution

Now we know that Equation (4.1) isn't the right form of gravitational force in general: it's only true over small distances compared to the separation of the two objects. So we can use it for objects falling a short distance toward the center of the Earth, since the falling object only travels a small fraction of the way from its starting point to the center of the Earth (where we imagine all the mass is concentrated). If we had two objects separated by any distance, the right equation of motion for a mass m at some distance from the origin where a fixed mass M is located would be

$$\mathbf{F}_g = -G\frac{Mm}{r^2}\hat{\mathbf{r}} \qquad (4.25)$$

where r is the distance from the origin. Again, thinking of just one dimension, it's easy to use Newton's Second Law to write down a more correct equation of motion

$$m\frac{d^2x}{dt^2} = -G\frac{Mm}{x^2} \qquad (4.26)$$

where x is the distance between the two masses. While the equation is simple to write down, it's not so easy to solve *this* equation by just integrating each side. (Try it! The problem is that we don't know what $x(t)$ on the right-hand side is, so we don't know how to integrate it.) However, we do know the starting value for the right-hand-side, we can discretize the derivatives and solve this one step at a time to build up the solution. But first, let's simplify our equations to make it easier to get a numerical solution.

4.3 The Power of Non-dimensionalizing

Instead of measuring distance in meters and time in seconds, it's often more convenient in a problem to measure them in terms of some "characteristic" size or time. For example, by defining a new variable $\rho = x/x_0$, where x_0 is a "characteristic size", all distances are now "pure numbers" (with no units) which measure lengths relative to x_0. With some care, it is usually possible to express *all* quantities in dimensionless form. Why would we want to do this? Well, there are several reasons, including not having to worry about units, but in the problem above, we have an intuition that the kind of function we get by solving the differential equation with the same initial conditions, but a different mass, say, would look very similar, just "scaled" in some way by the mass. Expressing the differential equation in a non-dimensional way lets us solve for the "shape" of the solution, and then re-scale it back to whatever mass we are interested in. This means we don't have to re-solve the equation each time we change the mass. Moreover, we generally find that the non-dimensional quantities like time and distance are helpful in defining the key physics of the problem, independent of the details of the solution of the differential equation. We will look more at this in the exercise at the end of this chapter.

Let's see how this works in practice. In the problem defined by Equation (4.26), the quantities with dimensions are position x, time t, mass M, and Newton's constant G. The only "characteristic" size anywhere in sight is x_0, the initial position, so let's try writing

$$\rho = \frac{x}{x_0} \tag{4.27}$$

where ρ, by definition, is dimensionless, and measures the fraction of the initial distance the particle has traveled. Substituting this in to Equation (4.26) gives

$$\frac{d^2\rho}{dt^2} = -\frac{GM}{x_0^3}\frac{1}{\rho^2} \tag{4.28}$$

Now, since the left-hand side has units of inverse time squared (ρ is dimensionless, but the dt^2 in the denominator carries the product of two times), we can conclude that the quantity

$$t_0 \equiv \sqrt{\frac{x_0^3}{GM}} \tag{4.29}$$

has dimensions of time. This allows us to also write a dimensionless time variable

$$\tau = t/t_0 \tag{4.30}$$

and then the resulting differential equation is very simple

$$\frac{d^2\rho}{d\tau^2} = -\frac{1}{\rho^2} \tag{4.31}$$

To see why this works out, it's helpful to just substitute our definitions from Equations (4.27) and (4.30) into the derivative:

$$\frac{d\rho}{d\tau} = \frac{d(x/x_0)}{d\tau} = \frac{1}{x_0}\frac{dx}{d\tau} = \frac{1}{x_0}\frac{dx}{d(t/t_0)} = \frac{t_0}{x_0}\frac{dx}{dt} \tag{4.32}$$

To tackle Equation (4.31) numerically, we'll use the computer, together with the trick in Equation (4.20). Let's write its two-first-order-equation form as

$$\frac{d\rho}{d\tau} = \xi(\tau) \tag{4.33}$$

$$\frac{d\xi}{d\tau} = -\frac{1}{\rho^2(\tau)} \tag{4.34}$$

Translating our initial conditions $x(0) = x_0$ and $\dot{x}(0) = 0$ into our new variables, we find

$$\rho_0 = \rho(0) = \frac{x(0)}{x_0} = \frac{x_0}{x_0} = 1 \tag{4.35}$$

$$\xi_0 = \frac{d\rho}{d\tau}(0) = \frac{t_0}{x_0}\frac{dx}{dt} = \frac{t_0}{x_0}\dot{x}(0) = 0 \tag{4.36}$$

where the last line used Equation (4.32).

So now the computer is going to solve Equations (4.33) and (4.34) by starting at ρ_0 and ξ_0, taking a small Δt and evaluating

$$\rho(\Delta t) = \xi_0 \Delta t \tag{4.37}$$

$$\xi(\Delta t) = -\frac{1}{\rho_0^2} \tag{4.38}$$

and in general, for n steps Δt, the solution will be

$$\rho_n \equiv \rho(n\Delta t) = \xi_{n-1}\Delta t \tag{4.39}$$

$$\xi_n \equiv \xi(n\Delta t) = -\frac{1}{\rho_{n-1}^2} \tag{4.40}$$

Notice how the solution at step n just depends on the solutions at the previous step $n - 1$. (In fact, most sophisticated differential equation solvers will use a more accurate approximation of the integral, but this is the conceptual idea.)

4.4 Summary

To recap, the key ideas are:
- Newton's Laws (and many other equations in physics) are written as differential equations for some function or functions.
- Fundamentally, we solve differential equations by integrating them.
- We can integrate a first-order differential equation numerically by directly appealing to the definition of integration
- For differential equations of second order (like Newton's Laws), we can transform to a system of first-order equations by introducing intermediate variables.
- Expressing the variables in terms of non-dimensional quantities allows us to simplify the form of the equations, and, as we will see, gain some insight into key parameters of the physical problem.

4.5 Exercise: Motion in a Constant Gravitational Field

In this exercise, we'll solve the problem of the motion of a particle of mass m located a distance x_0 from the origin, where a mass M is fixed. We'll do this for two different forces: a constant one, and one that looks like $1/r^2$ like the actual gravitational force. (See the figure below, which shows the configuration and the force acting on the mass m particle at $t = 0$.)

At first glance, this sounds like a problem in pure physics, but it's actually important for a key idea in astrophysics: gravitational collapse. Consider the material that was necessary to form the solar system. We think all of this material started as a cloud of gas and dust about 1 lightyear in radius, with a total mass a bit bigger than the Sun's current mass. The gravitational collapse of this cloud eventually resulted in the Sun and planets. So the problem of a particle falling under gravity actually represents an approximation to the question of how long it took that initial cloud of material to collapse, which gives an indication of how quickly stars and planetary systems can form. (There are lots of complications in reality, but this is a starting point.)

Along the way, we are going to
- Become familiar with a standard numerical method for integrating differential equations
- Solve the gravitational one-dimensional free-fall problem for a constant force and the correct $1/r^2$ form
- Gain a sense of the astronomical time, distance and velocity scales involved in free-fall for the case of star formation

First, let's solve the free-fall problem assuming the force is constant. For reasons that will become clear later, we will take the constant acceleration g to be the acceleration that a $1/r^2$ force would produce at the initial position of the particle. Thus, the equation we are going to solve is

$$m\frac{d^2x}{dt^2} = -mg = -m\frac{GM}{x_0^2} \tag{4.41}$$

Note that the mass of the falling object m cancels out, so we just have

$$\frac{d^2x}{dt^2} = -g \equiv -\frac{GM}{x_0^2} \tag{4.42}$$

where we're defining g by the above equation.

Our initial conditions are that

$$x(0) = x_0 \tag{4.43}$$

$$\frac{dx}{dt}(0) = \dot{x}(0) \equiv v_0 = 0 \tag{4.44}$$

We know the right answer with these conditions, namely

$$x(t) = x_0 - \frac{1}{2}gt^2 \tag{4.45}$$

so once we're done we can easily check if the numerical solutions are coming out right.

Recall that we defined a "characteristic time" t_c for this problem to be

$$t_c = \sqrt{\frac{x_0^3}{GM}} \tag{4.46}$$

The initial conditions are illustrated in Figure 4.1.

Question 1. Do this problem analytically (i.e., "on a piece of paper"). Write the differential equation above in dimensionless form using

$$x = x_0\rho \tag{4.47}$$

$$t = t_c\tau \tag{4.48}$$

where x_0 is the initial position and t_c is the characteristic time defined above.

Also write the *solution*

$$x(t) = x_0 - \frac{1}{2}gt^2 \tag{4.49}$$

in dimensionless form (in terms of ρ and τ).

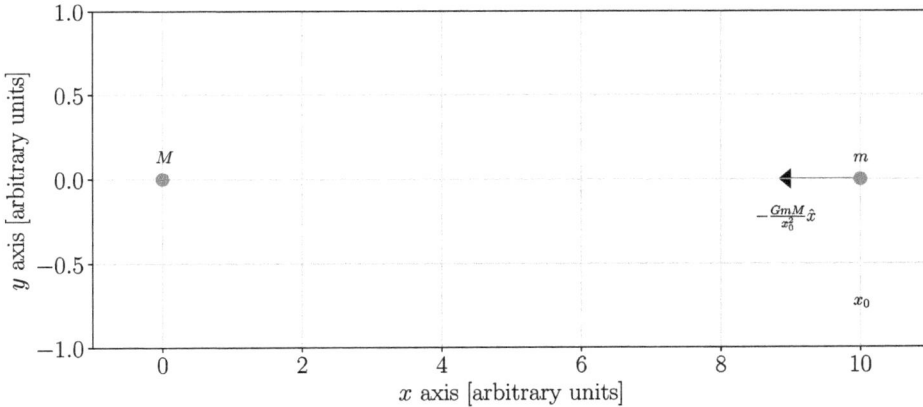

Figure 4.1. The initial conditions for the free-fall problem. We assume the mass M is fixed at the origin and the mass m begins initially at position x_0, with zero velocity, which will, however, be directed to the left as the particle begins to accelerate. Note the magnitude of the "gravitational constant" g.

Question 2. Again, do this problem analytically. Now use the trick of writing

$$\frac{d\rho}{d\tau} = \xi(\tau) \tag{4.50}$$

for some new function $\xi(\tau)$ to get two-first-order differential equations.

Question 3. Once we get the differential equation in this form, `scipy` has a fairly sophisticated differential equation solver in `scipy.integrate.solve_ivp`. (IVP stands for "initial value problem".) However, we need to write the math above as a function `solve_ivp` understands, where we calculate the quantities $\frac{d\rho}{d\tau}$ and $\frac{d\xi}{d\tau}$ given values of ρ, ξ, and τ. The template below shows what `solve_ivp` is expecting. Convert the following Markdown cell to code and fill in the question marks ??? below with the correct functions of rho, xi, and possibly tau. Then run it with the correct values for the initial conditions and verify you can access the solution ρ and τ from what is returned.

```
from scipy.integrate import solve_ivp

def EqMoConstForce(tau, vec):

    """ The system of first order equations for 1-D motion
    under a 1/r^2 force, using the
    unitless version derived in class.
    The components of vec are
        vec[0] = rho
        vec[1] = xi
    and tau is in the independent variable.

    Given values of rho and xi, this function returns the
    vector [drho, dxi]

    Appropriate for use with scipy.integrate.solve_ivp

    """

    rho = vec[0]
    xi = vec[1]

    drho = ???  # Replace with correct expression
    dxi = ???   # Replace with correct expression

    return [drho, dxi]
```

We will also need to supply the initial conditions, of course. By virtue of our non-dimensionalizing, we have

$$\rho(0) = \rho_0 = \frac{x_0}{x_0} = 1 \qquad (4.51)$$

$$\dot{\rho}(0) = \xi_0 = \frac{t_0}{x_0} v_0 = 0 \qquad (4.52)$$

We will also need to supply a range of times for which we want the solution, which is an interesting question. Since we are measuring the time in terms of the "characteristic time" t_0, let's try letting $0 < \tau < 1$.

Let's look at how we impart this information to solve_ivp.

```
# Define the initial conditions
rho0 = 1
xi0 = 0
init = [rho0, xi0]
# Define the range of desired tau
tau = np.linspace(0, 1, num=1000) # Note, no units!
"""
Call solve_ivp.  We give it the function we wrote to
    describe the system of differential
equations (nonDemFreeFall), the range of tau, the initial
    conditions, and the actual values
of tau where want the solution evaluated (t_eval).
"""
ff = solve_ivp(EqMoConstForce, [tau[0],tau[-1]], init,
    t_eval = tau)
```

`solve_ivp`

actually returns its result as an object, and we can access the `tau` values from `ff.t` (the "time" values), and $\rho(t)$ (position) as `ff.y[0,:]` and $\xi(t)$ (velocity) as `ff.y[1,:]`, because those are the zeroth and first quantities solved for in `EqMoConstForce`.

Question 4. Sketch on your piece of paper what you expect the plot of $\rho(\tau)$ and $\xi(\tau)$ to look like.

Make two plots, one of $\rho(\tau)$ (y-axis) versus τ (x-axis) and one of one of $\xi(\tau)$ (y-axis) versus τ (x-axis). Remember that ρ represents the position of the particle, and ξ its velocity. Remember to label your axes. Plot the exact solution (in non-dimensional form) on the plots. Is everything as expected? Comment.

Having put all this machinery together, it's fairly straightforward to change the force law. Let's solve the "exact" equation (where the form of the gravitational force is correct)[2]

$$m\frac{d^2x}{dt^2} = -G\frac{Mm}{x^2} \tag{4.53}$$

---

[2] I'm putting exact in quotes because, while this gravitational force law is definitely more correct, we find that that in physics, there's always more precision or different effects to include, so one shouldn't get too seduced by mathematical exactness when it's not necessary or warranted.

The notes showed its non-dimensional, first-order as form

$$\frac{d\rho}{d\tau} = \xi \tag{4.54}$$

$$\frac{\xi}{d\tau} = -\frac{1}{\rho^2} \tag{4.55}$$

with the same initial conditions as above.

We can go ahead and solve this, but we should also anticipate that there will be a problem at $x = 0$ corresponding to $\rho = 0$, since there the force is infinite.

Question 5. Use EqMoConstForce as a template to write a new function EqMoInvSqForce to solve this problem and calculate and solve_ivp to find a solution $\rho(\tau)$. Give your solution a different name than ff above. Make two plots, one of $\rho(\tau)$ (y-axis) versus τ (x-axis) and one of $\xi(\tau)$ (y-axis) versus τ (x-axis). Label your axes. Is the falling body at the origin, beyond it, or not yet to it in one "characteristic time"? Try tweaking the maximum value of τ to get the object close to $\rho = 0$.

Question 6. Now let's put units back on the problem. Convert ρ back to a physical distance and τ back into a time, assuming that x_0 is one lightyear and M is the mass of the Sun.

Plot $x(t)$ and $v(t)$ for the $1/r^2$ solution. Estimate how long does the object take to fall one lightyear under this force? How fast is it going when it reaches the origin? The time in seconds is not likely to be meaningful to you; also give it in years or a convenient multiple of years (thousands, millions, billions). The same for the velocity in meters per second; try km hr^{-1} or mph.

An Introduction to Astrophysics with Python
Stars and planets
James Aguirre

Chapter 5

Setting Up the "Two-body Problem"

Having looked at the numerical solution of the free-fall problem in one
dimension, we now consider the more general problem of the motion of a
system of two particles in three dimensions, interacting only via gravity,
the so-called "two-body problem". We describe simplifications that
reduce this problem to the motion of a single "fictitious" particle moving
only in a plane and conclude by writing down a set of equations
suitable for solving this simplified problem numerically.

5.1 The Two-body Problem

The basic problem we want to solve is that of two bodies of mass m_1 and m_2
interacting *only* under their mutual gravitational attraction. This is a very important
problem in astronomy, with obvious applications to the solar system but also to any
case in which we can assume that the gravitational forces relevant to the motion of
two bodies are dominated by their interaction only with one other.

We are going to start with the full set of differential equations for two bodies, and
our goal is to end up with a set of *first-order, dimensionless* differential equations to
describe the motion. We will find that there is a lot of simplification to be had in this
problem, so we're going to break it down as follows:

- First, we can change from a set of equations each describing one particle, to a
  single equation describing the motion of a single fictitious particle.
- Then, we show that instead of having to work in three dimensions, the motion
  actually happens all in a plane, so we only have to work in two dimensions.
- Finally, we make the equations dimensionless so we can get the essence of the
  solution, and re-scale back for any given masses and distances. This is
  important because the same physics will work for planets going around stars,
  for stars orbiting each other in pairs, for satellites going around planets, for a
  star orbiting the center of the Galaxy, etc, etc. The masses and distances

doi:10.1088/2514-3433/ade5f6ch5

change enormously between these problems, but the essential features of the orbital motion remain the same.

Our goal in writing first-order differential equations is so that we can solve the orbital motion on a computer (it turns out that solution for the trajectory does not have an analytic solution). In subsequent chapters, we will look at what we *can* learn analytically about the motion, and then put these two pieces together to get a fuller understanding of the ways that bodies move under gravitational forces.

5.2 The Full Set of Differential Equations

We begin by writing Newton's Laws for each body, where the position of each of the two bodies is represented by a vector $\mathbf{r}_i$, with three components, each describing the position of the body in (x, y, z) at time t:

$$m_1\ddot{\mathbf{r}}_1 = \frac{Gm_1m_2}{|\mathbf{r}_2 - \mathbf{r}_1|^2} \frac{\mathbf{r}_2 - \mathbf{r}_1}{|\mathbf{r}_2 - \mathbf{r}_1|} \tag{5.1}$$

$$m_2\ddot{\mathbf{r}}_2 = -\frac{Gm_1m_2}{|\mathbf{r}_2 - \mathbf{r}_1|^2} \frac{\mathbf{r}_2 - \mathbf{r}_1}{|\mathbf{r}_2 - \mathbf{r}_1|} \tag{5.2}$$

The vector $\mathbf{r}_2 - \mathbf{r}_1$ points from particle 1 to particle 2. (The minus sign in Equation (5.2) is so that the force points from 2 to 1.)

The end goal is to get $\mathbf{r}_1(t)$ and $\mathbf{r}_2(t)$ given the initial conditions $\mathbf{r}_1(t = 0)$, $\mathbf{r}_2(t = 0)$, $\dot{\mathbf{r}}_1(t = 0)$, and $\dot{\mathbf{r}}_2(t = 0)$. This is two second-order vector differential equations, which is hard. Let's start simple.

5.2.1 Kinematics

We need to set up some ground rules for how we describe motion, called kinematics, so we understand the meaning of Equations (5.1) and (5.2). We start by defining the position vector of a particle

$$\mathbf{r} = x(t)\hat{\mathbf{x}} + y(t)\hat{\mathbf{y}} + z(t)\hat{\mathbf{z}} \tag{5.3}$$

x in SI units is measured in meters. Then the velocity is the time derivative of **r**

$$\mathbf{v} = \frac{d\mathbf{r}}{dt} = \dot{\mathbf{r}} = \dot{x}(t)\hat{\mathbf{x}} + \dot{y}(t)\hat{\mathbf{y}} + \dot{z}(t)\hat{\mathbf{z}} \tag{5.4}$$

and the acceleration is the time derivative of velocity

$$\mathbf{a} = \frac{d\mathbf{v}}{dt} = \dot{\mathbf{v}} = \ddot{\mathbf{r}} = \ddot{x}(t)\hat{\mathbf{x}} + \ddot{y}(t)\hat{\mathbf{y}} + \ddot{z}(t)\hat{\mathbf{z}} \tag{5.5}$$

5.2.2 Center of Mass

The next thing to note is that, by Newton's Third Law, the gravitational forces are equal and opposite, and thus *sum to zero*. (This is a quite general statement about any system of particles that are not subject to some imposed "external force": the sum of all forces must be zero.) So if we add Equations (5.1) and (5.2), we get

$$m_1\ddot{\mathbf{r}}_1 + m_2\ddot{\mathbf{r}}_2 = 0 \tag{5.6}$$

or, pulling the derivative out,

$$\frac{d^2}{dt^2}(m_1\mathbf{r}_1 + m_2\mathbf{r}_2) = 0 \tag{5.7}$$

If we examine this quantity $m_1\mathbf{r}_1 + m_2\mathbf{r}_2$ we realize it's not quite a position vector (the units are wrong: kilograms × meters), but it's proportional to a vector we might have seen before, namely the vector pointing to the system's center of mass

$$\mathbf{R} \equiv \frac{m_1\mathbf{r}_1 + m_2\mathbf{r}_2}{M} \tag{5.8}$$

where we've defined the total mass $M = m_1 + m_2$. Since multiplying by $1/M$ doesn't change Equation (5.7), it immediately follows that the center of mass of the system obeys

$$\frac{d^2\mathbf{R}}{dt^2} = 0 \tag{5.9}$$

with the simple solution $\mathbf{R} = \mathbf{R}_0 + \mathbf{V}_0 t$ where $\mathbf{R}_0$ and $\mathbf{V}_0$ are constant vectors specifying the initial position and velocity of the center of mass.

Physically, this means that the center-of-mass motion is not very interesting, and we could choose to describe the motion in a frame where the center of mass is not moving, and then add its motion back in later. Equivalently, we can talk about the motion of the two bodies relative to each other, described by the vector $\mathbf{r}$, and the motion of the center of mass $\mathbf{R}$.

Making the simplifying assumption at $\mathbf{R} = 0$, the equation for the center of mass becomes

$$m_1\mathbf{r}_1 + m_2\mathbf{r}_2 = 0 \tag{5.10}$$

and the definition of the vector connecting the two masses is

$$\mathbf{r} = \mathbf{r}_2 - \mathbf{r}_1 \tag{5.11}$$

These two equations let us solve for $\mathbf{r}_1$ and $\mathbf{r}_2$ in terms of $\mathbf{r}$. Dividing by the total mass $M = m_1 + m_2$

$$\frac{m_1}{M}\mathbf{r}_1 + \frac{m_2}{M}\mathbf{r}_2 = 0 \tag{5.12}$$

so

$$\frac{m_1}{M}\mathbf{r}_1 + \frac{m_2}{M}(\mathbf{r}_1 + \mathbf{r}) = 0 \tag{5.13}$$

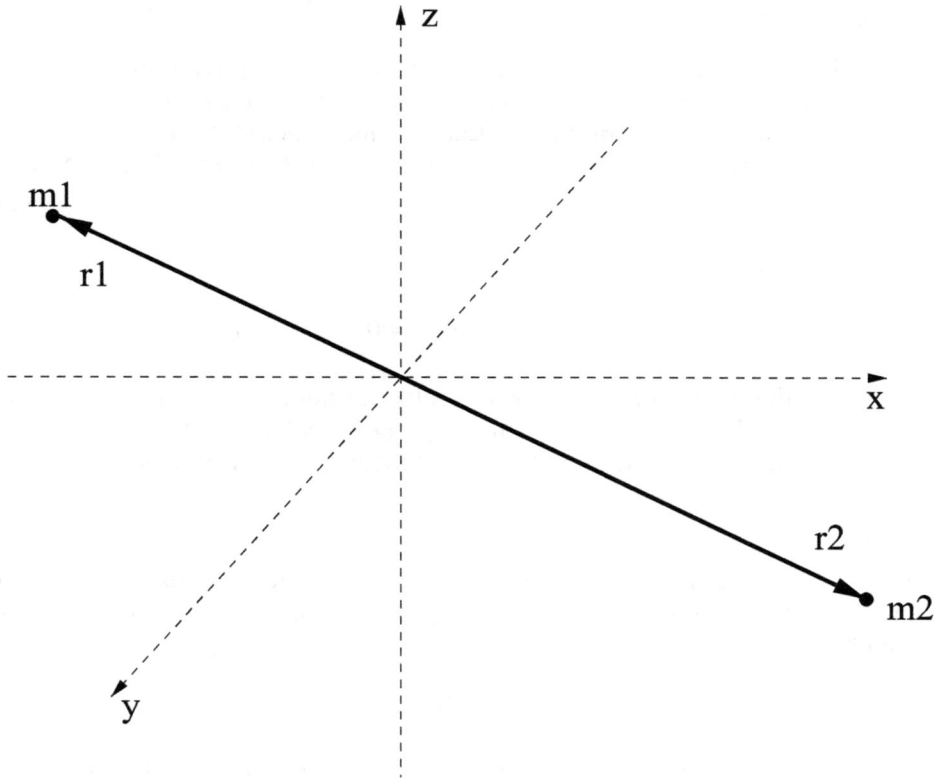

Figure 5.1. The geometry of the two-body problem, showing the location of the center of mass at the origin, and the orientations of the two vectors $\mathbf{r}_1$ and $\mathbf{r}_2$.

and we can write

$$\mathbf{r}_1 = -\frac{m_2}{M}\mathbf{r} \tag{5.14}$$

$$\mathbf{r}_2 = \frac{m_1}{M}\mathbf{r} \tag{5.15}$$

as shown in Figure 5.1.

So we've achieved the following: instead of the two vectors $\mathbf{r}_1$ and $\mathbf{r}_2$, we just need an equation for one vector $\mathbf{r}$, since the center of mass equation (for $\mathbf{R}$) was easy.

5.2.3 The Fictitious Problem

In Equation (5.9), we found the equation of motion of the center of mass position vector $\mathbf{R}$. We would now like to find the equation of motion for the position $\mathbf{r}$ describing the distance between the two bodies, which will turn out to define the motion of a fictitious particle of mass μ (to be defined shortly).

If we substitute the definitions

$$\mathbf{r} = \mathbf{r}_2 - \mathbf{r}_1$$

(Equation (5.11)) and

$$\mathbf{r}_1 = -\frac{m_2}{M}\mathbf{r}$$

(Equation (5.14)), into Equation (5.1), then we get

$$m_1\ddot{\mathbf{r}}_1 = -\frac{m_1 m_2}{M}\ddot{\mathbf{r}} \tag{5.16}$$

$$= \frac{Gm_1 m_2}{|\mathbf{r}_2 - \mathbf{r}_1|^2} \frac{\mathbf{r}_2 - \mathbf{r}_1}{|\mathbf{r}_2 - \mathbf{r}_1|} \tag{5.17}$$

$$= \frac{Gm_1 m_2}{|\mathbf{r}|^2} \frac{\mathbf{r}}{|\mathbf{r}|} \tag{5.18}$$

$$= \frac{Gm_1 m_2}{|\mathbf{r}|^2}\hat{\mathbf{r}} \tag{5.19}$$

Or, simplifying

$$\frac{m_1 m_2}{M}\ddot{\mathbf{r}} = -\frac{Gm_1 m_2}{r^2}\hat{\mathbf{r}} \tag{5.20}$$

This looks almost like the equation for a single particle, and if we make the definition of a quantity called the *reduced mass*

$$\mu \equiv \frac{m_1 m_2}{m_1 + m_2} = \frac{m_1 m_2}{M} \tag{5.21}$$

then Equation (5.20) looks like

$$\mu\ddot{\mathbf{r}} = -\frac{GM\mu}{r^2}\hat{\mathbf{r}} \tag{5.22}$$

so this looks like the equation of motion for the position $\mathbf{r}$ of a particle of mass μ under the gravitational force of a mass $M = m_1 + m_2$ fixed at the origin. Note that this is *not* the physical situation (both particles orbit the center of mass), but it is a very convenient mathematical description.

5.2.4 Conservation of Angular Momentum

Having gotten the problem down to just one 3D set of differential equations, let's look for one more simplification. We recall that the angular momentum of an object is defined by

$$\mathbf{L} = \mathbf{r} \times \mathbf{p} = m\mathbf{r} \times \mathbf{v} \tag{5.23}$$

where $\mathbf{p}$ is the linear momentum and $\times$ is the vector cross product. The total angular momentum of our system is the sum of the angular momenta of each particle:

$$\mathbf{L}_{\text{tot}} = m_1 \mathbf{r}_1 \times \mathbf{v}_1 + m_2 \mathbf{r}_2 \times \mathbf{v}_2 \qquad (5.24)$$

If we use Equations (5.14) and (5.15) in center of mass (CoM) coordinates, then we get

$$\begin{aligned} \mathbf{L} &= m_1 \mathbf{r}_1 \times \mathbf{v}_1 + m_2 \mathbf{r}_2 \times \mathbf{v}_2 \\ &= -\frac{m_1 m_2}{M} \mathbf{r} \times \mathbf{v}_1 + \frac{m_1 m_2}{M} \mathbf{r} \times \mathbf{v}_2 \\ &\equiv \mu \mathbf{r} \times (\mathbf{v}_2 - \mathbf{v}_1) \end{aligned}$$

If we also notice that

$$\dot{\mathbf{r}} = \dot{\mathbf{r}}_2 - \dot{\mathbf{r}}_1 \equiv \mathbf{v} \qquad (5.25)$$

and we define $\mathbf{p} = \mu \mathbf{v}$, then the total angular momentum looks like

$$\mathbf{L}_{\text{tot}} = \mathbf{r} \times \mathbf{p} \qquad (5.26)$$

which is the angular momentum for a particle of mass μ moving around a point at a distance $\mathbf{r}$. (This is, in fact, what we should expect from using CoM coordinates!)

There is a particularly important property of $\mathbf{L}$. Consider

$$\frac{d\mathbf{L}}{dt} = \frac{d}{dt}(\mathbf{r} \times \mathbf{p}) = \dot{\mathbf{r}} \times \mathbf{p} + \mathbf{r} \times \dot{\mathbf{p}} \qquad (5.27)$$

But $\dot{\mathbf{r}} = \mathbf{v}$, which is parallel to $\mathbf{p} = m\mathbf{v}$, so the first cross product is zero, and by Newton's Second Law, $\dot{\mathbf{p}} = \mathbf{F}$, and $\mathbf{F}$ is parallel to the vector $\mathbf{r}$ between the two particles. So that term is zero as well. Note that this would be true for *any* force $\mathbf{F}$ which is only dependent on the direction between particles (not just the gravitational force we're thinking of here). Thus, we have

$$\boxed{\frac{d\mathbf{L}}{dt} = 0} \qquad (5.28)$$

or $\mathbf{L} = $ constant, both in magnitude and in direction. Since $\mathbf{r}$ and $\mathbf{p}$ are always perpendicular to $\mathbf{L}$ (by definition), and the direction of $\mathbf{L}$ is fixed, all of the motion must occur in the plane perpendicular to $\mathbf{L}$. So we can choose the object to be in the (x, y)-plane.

To review: we now know that the important dynamics of the problem depend only on the vector between the two particles $\mathbf{r}$, and actually only on the x and y components of $\mathbf{r}$.

5.3 Setting Up the Problem for Numerical Solution

Now we know we only care about the x and y components of $\mathbf{r}$. Thus

$$|\mathbf{r}|^2 = x^2 + y^2 \qquad (5.29)$$

and the vector $\hat{\mathbf{r}}$ points from the origin to a point (x, y) and has unit length:

$$\hat{\mathbf{r}} = \frac{1}{\sqrt{x^2 + y^2}}(x\hat{\mathbf{x}} + y\hat{\mathbf{y}}) \tag{5.30}$$

Then we have

$$\ddot{x}(t)\hat{\mathbf{x}} + \ddot{y}(t)\hat{\mathbf{y}} = -\frac{GM}{x^2 + y^2}\frac{1}{\sqrt{x^2 + y^2}}(x\hat{\mathbf{x}} + y\hat{\mathbf{y}}) \tag{5.31}$$

$$= -\frac{GM}{(x^2 + y^2)^{3/2}}(x\hat{\mathbf{x}} + y\hat{\mathbf{y}}) \tag{5.32}$$

or in matrix form

$$\begin{bmatrix} \ddot{x}(t) \\ \ddot{y}(t) \end{bmatrix} = -\frac{GM}{(x^2 + y^2)^{3/2}}\begin{bmatrix} x \\ y \end{bmatrix} \tag{5.33}$$

We can non-dimensionalize this by taking a characteristic distance to be r_c and

$$X = \frac{x}{r_c} \tag{5.34}$$

$$Y = \frac{y}{r_c} \tag{5.35}$$

and, as in the free-fall problem, we set $\tau = t/t_c$ with

$$t_c = \sqrt{\frac{r_c^3}{GM}} \tag{5.36}$$

leading to

$$\begin{bmatrix} \frac{d2X}{d\tau 2} \\ \frac{d2Y}{d\tau 2} \end{bmatrix} = \begin{bmatrix} \ddot{X} \\ \ddot{Y} \end{bmatrix} = -\frac{1}{(X^2 + Y^2)^{3/2}}\begin{bmatrix} X \\ Y \end{bmatrix} \tag{5.37}$$

so that the dot notation now indicates derivatives with respect to τ. Now defining

$$V_X = \frac{dX}{d\tau} \equiv \dot{X} \tag{5.38}$$

$$V_Y = \frac{dY}{d\tau} \equiv \dot{Y} \tag{5.39}$$

then we can write this as the first-order system of equations

$$
\begin{bmatrix} \dot{X} \\ \dot{V_X} \\ \dot{Y} \\ \dot{V_Y} \end{bmatrix} = \begin{bmatrix} V_X \\ -(X^2 + Y^2)^{-3/2}X \\ V_Y \\ -(X^2 + Y^2)^{-3/2}Y \end{bmatrix} \tag{5.40}
$$

5.4 Exercise: Solving the Two-body Problem Numerically

In this exercise, we want to
- Understand the formulation of the two-body problem as a (dimensionless) differential equation
- Obtain solutions for orbits from a numerical differential equation solver
- Gain some intuition for the behavior of the orbit as a function of the initial conditions
- Understand the behavior of the position and velocity as a function of time

5.4.1 Background

Recall that by using the center of mass coordinate system and conservation of angular momentum, we've reduced the full set of differential equations for two bodies interacting gravitationally (two 3D second-order differential equations) to four first-order equations describing the motion of a single fictitious particle of mass $\mu = m_1 m_2 / (m_1 + m_2)$ moving around the center of mass of the system in two dimensions.

Written in dimensionless form, this set of differential equations is

$$
\begin{bmatrix} \dot{X} \\ \dot{V_X} \\ \dot{Y} \\ \dot{V_Y} \end{bmatrix} = \begin{bmatrix} V_X \\ -(X^2 + Y^2)^{-3/2}X \\ V_Y \\ -(X^2 + Y^2)^{-3/2}Y \end{bmatrix} \tag{5.41}
$$

For the case where one mass is much less than the other (for example, a planet going around the Sun), the solution of the fictitious problem is very close to the lighter mass orbiting the position of the heavier one, which is what Kepler said. Recall that we can always translate from the fictitious problem back to the actual positions of the two particles separately, but for now we'll stay in the limit $m_1 \gg m_2$ and think of the solution to the above equations as being the orbit of the lighter particle.

5.4.2 Setting It Up for the Computer

We're now going to take Equation (5.40) straight to the computer and see what kind of solutions emerge. Recall that the differential equation solver `solve_ivp` needs three things:

- an implementation of the equations above as a function that it can execute
- the initial conditions at which to start the solution, and
- the time points at which the solution is desired.

Let's take each of these in turn. The function is defined below.

```python
def two_body_2d_cart_eq_mo(tau, vec):

    """
    The nondimensional two-body problem in Cartesian
    coordinates.

    The order of the components of the input vector Vec is
    X, dX/dtau, Y, dY/dtau.

    Appropriate for use with scipy.integrate.solve_ivp.
    Also backwards compatible with scipy.integrate.odeint
    using tfirst=True
    """

    # Rename the values in the input vector Vec to be more
    readable by humans
    X = vec[0]
    VX = vec[1]
    Y = vec[2]
    VY = vec[3]

    # Calculate r-squared
    r2 = np.power(X, 2) + np.power(Y, 2)

    # Initialize the right hand side
    dvecdtau = [0,0,0,0]

    dvecdtau[0] = VX
    dvecdtau[1] = -1./np.power(r2, 3./2.) * X
    dvecdtau[2] = VY
    dvecdtau[3] = -1./np.power(r2, 3./2.) * Y

    return dvecdtau
```

The initial conditions are given as a 4-element list (since we have four equations), as, for example

```
init = [1,0,0,1]
```

In our dimensionless units, $0 \leqslant \tau \leqslant 6\pi$ works reasonably well

```
tau = np.linspace(0,6.*np.pi,num=50000)
```

so we can invoke the solver as

```
orbit = solve_ivp(two_body_2d_cart_eq_mo, [tau[0], tau
    [-1]],
                  init, t_eval = tau,
                  method='LSODA', rtol=1e-6)
```

Note that there are a couple of extra arguments to solve_ivp which are of the form keyword=value, in this case method and rtol. Such "keywords" are optional arguments, but in this case they are necessary to get solve_ivp to do a good job with this particular problem, which was not necessary for Exercise 4.5, Question 3.

solve_ivp returns its solutions for the positions and velocities in the variable orbit, which is an *object*, a very flexible variable type that can depend on context. In order to extract the values for the times at which the solution was calculated, as well as the positions and velocities, you can use the syntax

```
tau = orbit.t
X = orbit.y[0,:]
VY = orbit.y[1,:]
Y = orbit.y[2,:]
VY = orbit.y[3,:]
```

Note the way this works: there is data named y associated with the orbit object, which is accessed using orbit.y; we can see that it is a numpy array, and has shape

```
orbit.y.shape
(4, 50000)
```

Note the notation for extracting the 0th dimension along the first axis (the X variable), and all the points along the second axis: [0, :], and similarly for the other variables.

The following function wraps up all this and returns everything as a dictionary to make it a little easier to run multiple solutions with different initial conditions.

```
def two_body_2d_cart(tau, init):

    orbit = solve_ivp(two_body_2d_cart_eq_mo, [tau[0], tau
    [-1]],
                      init, t_eval = tau,
                      method='LSODA', rtol=1e-6)

    return {'X': orbit.y[0,:],
            'VX': orbit.y[1,:],
            'Y': orbit.y[2,:],
            'VY': orbit.y[3,:],
            'tau': orbit.t}
```

Question 1. We are going to consider six different cases of initial conditions. For future use, you will want to save all your solutions as separate variables so that you can refer back to the solutions in the following questions (say, orbit1, orbit2, etc).

Obtain solutions $X(\tau)$, $Y(\tau)$ for the two-body problem for the initial conditions given below. The first five correspond to

$$X_0 = 1 \tag{5.42}$$

$$Y_0 = 0 \tag{5.43}$$

$$V_{X0} = 0 \tag{5.44}$$

for five different initial velocities $V_{Y0} = 0.6, 1, 1.23, \sqrt{2}, 2$, as well as one with $V_{X0} = 1$, $V_{Y0} = 0.716$.

Plot $X(\tau)$ versus $Y(\tau)$ to show the shape of the orbit. Indicate the origin with a dot, and label the axes (these are dimensionless units, but let's make it explicit).

Question 2. Answer the following qualitative questions:
- For which initial conditions is the orbit clearly elliptical?
- Is any orbit circular?
- Describe qualitatively what is happening in the cases that are *not* ellipses.
- How would you describe what is different about the last initial condition (where V_{X0} and V_{Y0} are both non-zero)?

Question 3. Note that you are plotting the position in the (X, Y)-plane, but not explicitly plotting τ. Where in (X, Y) is $\tau = 0$? As τ increases, in what direction is the particle moving? Make a plot with the orbits just for the first 5000 points to show this more clearly.

Note that you extract or "index" the first 5000 values like this: orbits['X'] [0:5000]

Question 4. For this question, just consider the initial conditions that produce circular or elliptical orbits. Recall that by construction, $(0,0)$ is the location of the center of mass (or of the larger mass in the case that one is much larger than the other). For which values of V_{Y0} is the initial position at perihelion? For which values at aphelion? Describe your reasoning for determining this, and feel free to make reference to the plots you made.

Question 5. For the orbits that are ellipses we can measure the semimajor axis by `(x.max()-x.min())/2`. (again, because of the way I set the problem up). For the first three initial conditions above, what is the trend of semimajor axis as V_{Y0} increases? Make a plot (with three points) of semimajor axis versus V_{Y0}.

Question 6. For the case $X_0 = 1$, $V_{Y0} = 0.6$, plot $\rho = \sqrt{X(\tau)^2 + Y(\tau)^2}$ versus τ. Describe the key features of $\rho(\tau)$. Where in the orbit does ρ reach its minimum and maximum values? Remember, you can calculate the variable ρ using numpy functions; no for loops required.

Question 7. Again, for $X_0 = 1$, $V_{Y0} = 0.6$, make a plot of $|V| = \sqrt{V_X^2 + V_Y^2}$. Compare your plot to that of Question 6. Where is the planet located when it is moving the fastest? Slowest?

5.5 Study Questions

1. Suppose a spacecraft is orbiting at an altitude of 575 km above the surface of the Earth. Remember that gravitational forces and orbital distances are calculated between the center of mass of objects. You will need to look up or use astropy's values for the mass and the radius of the Earth to solve this problem.

    (a) What is the period of the spacecraft's orbit?

    (b) Assuming the orbit is a circle, how fast is the spacecraft traveling? Write down an expression for the speed before plugging in numbers. (Hints: Here we're just using the basic definition of speed as distance/time.)

    (c) This spacecraft is orbiting at an altitude higher than the space station. But suppose the crew wants to go *even higher*. Would this higher orbit have them traveling faster or slower? I would like to see a symbolic answer and reasoning, but you may find it helpful to check your thinking by plugging in to your answer to Part 1b.

(d) Suppose the spacecraft fires its rockets in such a way as to make its motion relative to the Earth zero, and it then starts to free fall straight down. Calculate the ratio of the acceleration due to gravity at its altitude of 575 km to what it would be on the ground. Using this ratio and the results of the exercise in Section 4.5, explain whether you need to consider a $1/r^2$ force law for calculating an accurate answer to the question of how long it would take spacecraft to fall this distance or whether a constant force is sufficient.

(e) Calculate how long it would take for the spacecraft to fall from 575 km to the surface of the Earth (ignoring the effects of the atmosphere or, you know, those essential parachutes). How long would it take to fall the full distance to the center of the Earth (assuming all the Earth's mass was concentrated at the center)? How different is that answer if you use a constant force versus a $1/r^2$ force?

2. You may have heard the astronomical unit (AU) defined as the "average" distance of the Earth from the Sun. However, when averaging, one has to define what samples one is averaging over.[1] Here we are going to look at two possible definitions for the average, first by looking at how far away the points of an ellipse are from one focus, averaging over $r(\theta)$ for a bunch of equally spaced samples in $\Delta\theta$, and then averaging over $r(t)$ for an actual orbit (with equally spaced samples in Δt). Recall that the average value of a function $f(x)$ over the interval a to b is given by

$$\langle f \rangle_x \equiv \frac{1}{b-a} \int_a^b f(x)dx$$

The following values of initial position X_0 and V_{Y0} with $Y_0 = 0$ and $V_{X0} = 0$ for our non-dimensional version of the problem will produce ellipses with a semimajor axis of 1 and an eccentricity as given.

e	X_0	V_{Y0}
0.0	1.0	1.0000
0.2	0.8	1.2247
0.4	0.6	1.5275
0.6	0.4	2.0000
0.8	0.2	3.0000

(a) Use the code in Exercise I.4 to calculate X- and Y-coordinates for all 5 orbits above with the given initial conditions. You will need to use a range of $0 \leqslant \tau \leqslant 2\pi$. Make sure you evaluate this for a large number of equally spaced time points. Plot all 5 orbits on the same set of axes. Calculate $r(\tau) = \sqrt{X^2 + Y^2}$ and plot this versus τ for all 5 orbits on a separate plot.

---

[1] In point of fact, the definition of the AU actually is fairly subtle and is no longer defined as an average distance. See https://en.wikipedia.org/wiki/Astronomical_unit.

(b) Use your previous code (from class) for calculating the equation of an ellipse in polar coordinates to calculate $r(\theta)$ for ellipses with the same a and e as the ellipses in Part 2a. Plot this on the same set of axes (you will need to convert to Cartesian coordinates!) as the ellipses in Part 2a. Plot $r(\theta)$ on the same set of axes as $r(\tau)$ above. (Because of how we're measuring "time", both θ and τ run from 0 to 2π.) Make sure the shape of the orbits in (X, Y) match for orbits with the same eccentricity before proceeding. It is OK if the shapes of $r(\tau)$ and $r(\theta)$ do not match for the same eccentricity.

(c) Use what we learned about numerical integration in Exercise I.5 to calculate for each of the orbits in 2a the quantity[2]

$$\langle r \rangle_\tau (e) = \frac{1}{T} \int_0^T r(\tau; e) d\tau$$

where T is the period of the orbit. Note that $r(\tau) = r(\tau; e)$ is also a function[3] of the eccentricity e. Plot this time-averaged distance as a function of eccentricity e.

(d) Use what we learned about numerical integration in Exercise I.5 to calculate for each of the orbits in 2a the quantity

$$\langle r \rangle_\theta (e) = \frac{1}{2\pi} \int_0^{2\pi} r(\theta; e) d\theta$$

Add the plot of this average distance to the plot of $\langle r \rangle_\tau$ versus e.

(e) Interpret, in words, what the two averages mean and why they disagree for most values of e, and explain why they *do* agree for a certain value of e.

---

[2] The sense of the notation $\langle r \rangle_\tau (e)$ is: "average $\langle\rangle$ the function r over the independent variable τ; the result is a function of the value of e." Note that the subscript τ on the left hand side is a reminder of the variable that was integrated over on the right, but the final answer—due to the integration—is *not* a function of τ.
[3] The sense of the notation $r(\tau; e)$ is "r is a function of the independent variable τ, but also depends on the value of the *parameter e*."

An Introduction to Astrophysics with Python
Stars and planets
James Aguirre

Chapter 6

Energetics in the Two-body Problem

Having solved for the motion in the two-body problem, we now look at how to extract more general features of the behavior by considering "conservation laws", specifically of energy and angular momentum. Using this language, we are able to derive and formulate Kepler's Laws not as arbitrary rules promulgated by fiat, but as necessary consequences of Newton's Laws and with numerical values calculable from the energy and angular momentum of the situation.

6.1 Energy and Newton's Laws

Recall that Newton's Second Law defines force as a mass times an acceleration, or, equivalently, as the time derivative of a momentum:

$$\mathbf{F} = m\mathbf{a} = m\dot{\mathbf{v}} = \frac{d(m\mathbf{v})}{dt} = \dot{\mathbf{p}} \tag{6.1}$$

The work or energy is defined as

$$E = \int_{\text{path}} \mathbf{F} \cdot d\mathbf{r} \tag{6.2}$$

where the integral is taken along a particular path, that is, a particular choice of $\mathbf{r}(t)$. The units of energy are force times distance k gm^2 s^{-2} or Nm, also known as Joules [J].

A form of energy you're probably familiar with is *kinetic energy*, which actually follows from applying the definition in Equation (6.2) to Newton's Second law, with the "path" being the path followed by the particle $\mathbf{r}(t)$ according to the law of motion

$$E_{\text{kin}} = \int_{\text{path}} \mathbf{F} \cdot d\mathbf{r} = \int_{\text{path}} m\frac{d^2\mathbf{r}}{dt^2} \cdot d\mathbf{r} = \int_0^t m\frac{d^2\mathbf{r}}{dt'^2} \cdot \frac{d\mathbf{r}}{dt'} \, dt' \tag{6.3}$$

doi:10.1088/2514-3433/ade5f6ch6

Now we can cleverly re-write the product of the acceleration and the velocity above by noting that

$$\frac{d(\mathbf{v} \cdot \mathbf{v})}{dt} = \mathbf{v} \cdot \frac{d\mathbf{v}}{dt} + \frac{d\mathbf{v}}{dt} \cdot \mathbf{v} = 2\mathbf{v} \cdot \frac{d\mathbf{v}}{dt} = 2\frac{d^2\mathbf{r}}{dt^2} \cdot \frac{d\mathbf{r}}{dt} \tag{6.4}$$

by the product rule, so

$$E_{\text{kin}} = \frac{1}{2}m \int_0^t \frac{d(\mathbf{v} \cdot \mathbf{v})}{dt'} \, dt' = \frac{1}{2}mv^2 \tag{6.5}$$

We can write the kinetic energy in several different ways, including as follows

$$\boxed{E_{\text{kin}} = \frac{1}{2}mv^2 = \frac{1}{2}m \, |\dot{\mathbf{r}}|^2} \tag{6.6}$$

Note that only *changes* in energy are meaningful, that is, the difference in energy at two different points along the path; this follows naturally from the definition of energy as an integral in Equation (6.2). Note that in general, E will depend on the path taken (think of a frictional force), but in the case that it doesn't, we can define a potential energy Φ such that

$$\mathbf{F} = -\nabla \Phi \tag{6.7}$$

where Φ is some function of position and ∇ is the gradient of that function; explicitly in Cartesian coordinates, cylindrical, and spherical coordinates:

$$\nabla \Phi = \frac{\partial \Phi}{\partial x}\hat{\mathbf{x}} + \frac{\partial \Phi}{\partial y}\hat{\mathbf{y}} + \frac{\partial \Phi}{\partial z}\hat{\mathbf{z}} \tag{6.8}$$

$$= \frac{\partial \Phi}{\partial r}\hat{\mathbf{r}} + \frac{1}{r}\frac{\partial \Phi}{\partial \theta}\hat{\boldsymbol{\theta}} + \frac{\partial \Phi}{\partial z}\hat{\mathbf{z}} \tag{6.9}$$

$$= \frac{\partial \Phi}{\partial r}\hat{\mathbf{r}} + \frac{1}{r}\frac{\partial \Phi}{\partial \theta}\hat{\boldsymbol{\theta}} + \frac{1}{r \sin \theta}\frac{\partial \Phi}{\partial \phi}\hat{\boldsymbol{\phi}} \tag{6.10}$$

(Note that although we use the same variable r and θ in cylindrical coordinates, the meaning is different; $r = \sqrt{x^2 + y^2}$ in cylindrical and $r = \sqrt{x^2 + y^2 + z^2}$ in spherical; similarly θ is in the (x, y)-plane in cylindrical, but ϕ is that same angle in spherical.)

This path-independence turns out to be true for the gravitational force, and going back to the definition Equation (6.2), we can integrate along a path that brings a mass m_2 in from infinity to some distance r from the mass m_1 at the origin as

$$E = \int_\infty^r \mathbf{F}(\mathbf{r}') \cdot d\mathbf{r}' \tag{6.11}$$

$$= \int_\infty^r \frac{Gm_1m_2}{r'^2}\hat{\mathbf{r}}' \cdot d\mathbf{r}' \tag{6.12}$$

$$= \int_\infty^r \frac{Gm_1 m_2}{r'^2} dr' \tag{6.13}$$

$$= -\frac{Gm_1 m_2}{r'} \bigg|_\infty^r \tag{6.14}$$

$$= -\frac{Gm_1 m_2}{r} \tag{6.15}$$

Note the minus sign; this is not a mistake. This minus sign follows naturally from the math, but we want to keep in mind what it means physically. Specifically, we've defined the gravitational potential energy to be zero as $r \to \infty$ (which makes sense, since the force is also zero at infinite distance). But in keeping with our notion that *differences* in energy are what's important, we're measuring relative to this zero energy, so that adding kinetic energy *increases* the total energy and adding potential energy *decreases* the total energy. We can write the gravitational potential energy in several ways:

$$\boxed{E_{\text{grav.pot}} \equiv \Phi = -\frac{Gm_1 m_2}{r} = -\frac{Gm_1 m_2}{|\mathbf{r}|}} \tag{6.16}$$

Note that, if we go back to Equation (6.7), we can verify this does indeed produce the proper force (in spherical coordinates):

$$\mathbf{F} = -\nabla \Phi = \frac{\partial}{\partial r}\left(-\frac{Gm_1 m_2}{r}\right)\hat{\mathbf{r}} = -\frac{Gm_1 m_2}{r^2}\hat{\mathbf{r}} \tag{6.17}$$

6.1.1 The Two-body Problem in Terms of Energy

Recall that after a great deal of simplification, we had re-written the two-body problem (compare Equation (5.22)) as

$$\mu\ddot{\mathbf{r}} = -\frac{GM\mu}{r^2}\hat{\mathbf{r}} \tag{6.18}$$

Mathematically, we reduced the original problem to a "fictitious" one in which the vector $\mathbf{r}$ describes the motion of a particle of mass μ around a *fixed* mass M. Notice that the force points from the location of the particle μ back toward the origin.

The energy for this problem is now easy to write down as the sum of the kinetic and gravitational potential energies

$$\boxed{E = \frac{1}{2}\mu |\dot{\mathbf{r}}|^2 - \frac{GM\mu}{|\mathbf{r}|}} \tag{6.19}$$

and we can use the planar motion to re-write this in polar coordinates (recalling Equation (3.25) for $|\dot{\mathbf{r}}|^2$ in polar coordinates), giving

$$E = \frac{1}{2}\mu(\dot{r}^2 + r^2\dot{\theta}^2) - \frac{GM\mu}{r} \tag{6.20}$$

Now, this has simplified a lot, and it would allow us to us write the energy solely in terms of r and $\dot{r}$, if only we could eliminate $\dot{\theta}$ from the expression above. Fortunately, we note that the angular momentum L is a constant and can be written as

$$L = \mu|\mathbf{r} \times \mathbf{v}| = \mu|r\hat{\mathbf{r}} \times (\dot{r}\hat{\mathbf{r}} + r\dot{\theta}\hat{\boldsymbol{\theta}})| = \mu|r^2\dot{\theta}\hat{\mathbf{z}}| \tag{6.21}$$

or explicitly

$$\boxed{L = \mu r^2 \dot{\theta}} \tag{6.22}$$

We're going to use this again in Section 6.2.2 in our discussion of Kepler's Second Law.
Using the relation for L, we can now re-write $\dot{\theta}$ as

$$\dot{\theta} = \frac{L}{\mu r^2} \tag{6.23}$$

and substituting this back into Equation (6.20), we get

$$\boxed{E = \frac{1}{2}\mu\dot{r}^2 + \frac{L^2}{2\mu r^2} - \frac{GM\mu}{r}} \tag{6.24}$$

We now have an expression for the energy of the two-body problem, written in terms *only* of the scalar function $r(t)$ which gives the separation between the two bodies, its time derivative, the total mass of the two bodies M, the reduced mass μ, and the angular momentum of the orbit L.

One thing that is not obvious about Equation (6.24) is that the total energy is *conserved*, meaning that quantity does not change with time. The individual terms in that equation, and indeed the kinetic and potential energy separately, *can* and do change with time, but the total is constant. This is what makes energy such a useful concept.

What is also not obvious from Equation (6.24) is that, depending on the situation, the total energy can be negative, zero, or positive. The angular momentum is always $L \geqslant 0$. The reason why energy is conserved and the behavior of the equations of motion when the energy takes on different values are explored further in Chapter 8.

6.2 Deriving Kepler's Laws

In the following, we are going to make use of the fact that $E < 0$ for the case of elliptical orbits, which is shown a bit more explicitly in Chapter 8.

6.2.1 Kepler's First Law

Recall that Kepler's First Law is that planets' orbital motion is in the shape of an ellipse with the Sun at one focus. So we should expect to find that the curve traced out by $(r(t), \theta(t))$ is, in fact, an ellipse. Actually solving for an explicit form of $r(t)$ that satisfies Equation (6.24) isn't possible. (This is why we get our orbits as a function of time numerically.) But if we want to figure out the *shape* of the orbit, we really just want to find $r(\theta)$. To do this, we first solve Equation (6.24) for $\dot{r}(t)$:

$$\dot{r} = \sqrt{\frac{2}{\mu}\left(E + \frac{GM\mu}{r}\right) - \frac{L^2}{2\mu r^2}} \tag{6.25}$$

and changing variables from dr/dt to $d\theta/dr$, to find

$$\frac{d\theta}{dr} = \frac{d\theta}{dt}\frac{dt}{dr} = \frac{\dot{\theta}}{\dot{r}} = \frac{L}{\mu r^2 \dot{r}} \tag{6.26}$$

where the last equality follows from our relation for angular momentum, Equation (6.22). Plugging Equation (6.25) in for $\dot{r}$ gives

$$\frac{d\theta}{dr} = \frac{L}{\mu r^2}\left(\frac{2}{\mu}\left(E + \frac{GM\mu}{r}\right) - \frac{L^2}{2\mu r^2}\right)^{-1/2} \tag{6.27}$$

Since the right-hand side now only depends on r and constants, it can be integrated[1] to give $\theta(r)$, which is then inverted to get

$$\boxed{r(\theta) = \frac{L^2/(GM\mu^2)}{1 + \sqrt{1 + \frac{2EL^2}{G^2M^2\mu^3}}\cos\theta}} \tag{6.28}$$

which is indeed the equation for an ellipse. Comparing this to

$$r(\theta) = \frac{a(1 - e^2)}{1 + e\cos\theta} \tag{6.29}$$

we can now relate the physical parameters E, L, M, and μ to the geometrical parameters a and e. We can read off e directly by looking at the denominator:

$$e = \sqrt{1 + \frac{2EL^2}{G^2M^2\mu^3}} \tag{6.30}$$

or

$$e^2 = 1 + \frac{2EL^2}{G^2M^2\mu^3} \tag{6.31}$$

Note that since E is *negative*, this allows $e = 0$ (and in fact imposes a constraint on the relation between E and L for circular orbits). We could also write this as

$$\boxed{e^2 = 1 - \frac{2|E|L^2}{G^2M^2\mu^3}} \tag{6.32}$$

---

[1] Using a number of clever tricks, not shown here.

We can also read off

$$a(1 - e^2) = \frac{L^2}{GM\mu^2} \tag{6.33}$$

$$= -a\frac{2EL^2}{G^2M^2\mu^3} \tag{6.34}$$

or

$$a = \frac{-GM\mu}{2E} = \frac{-Gm_1m_2}{2E} \tag{6.35}$$

Since, again, we know $E < 0$, this is

$$\boxed{a = \frac{GM\mu}{2|E|}} \tag{6.36}$$

6.2.2 Kepler's Second Law

Looking back at our statement of Kepler's Second Law in terms of equal area in equal time, Equation (3.32) we found that for elliptical orbits

$$\frac{dA}{dt} = \frac{1}{2}r^2\frac{d\theta}{dt} = \frac{1}{2}r^2\dot{\theta} = \text{constant} \tag{6.37}$$

and recalling that $L = \mu r^2 \dot{\theta}$ is also constant, we can now see that explicitly the Kepler's Second Law constant must be

$$\boxed{\frac{dA}{dt} = \frac{1}{2}\frac{L}{\mu}} \tag{6.38}$$

Thus, in essence, Kepler's Second Law is a statement of the conservation of angular momentum.

6.2.3 Kepler's Third Law

Kepler's Third Law now follows from the Second, which we re-write as:

$$dt = \frac{2\mu}{L}dA \tag{6.39}$$

where we now want to integrate the time over one period

$$\int_0^P dt = P = \int_0^A \frac{2\mu}{L}dA = \frac{2\mu}{L}A = \frac{2\mu}{L}\pi ab \tag{6.40}$$

which follows since we know the area of the ellipse. Thus

$$P = \frac{2\mu}{L}\pi ab \tag{6.41}$$

and we can eliminate b and L using

$$b^2 = a^2(1 - e^2) \tag{6.42}$$

and

$$a(1 - e^2) = \frac{L^2}{GM\mu^2} \tag{6.43}$$

so that

$$b^2 = \frac{aL^2}{GM\mu^2} \tag{6.44}$$

or

$$\frac{b^2}{L^2} = \frac{a}{GM\mu^2} \tag{6.45}$$

This gives

$$P = 2\pi\mu a\sqrt{\frac{a}{GM\mu^2}} = 2\pi\sqrt{\frac{a^3}{GM}} \tag{6.46}$$

One remarkable thing about this relation is that we've already seen that the combination $\sqrt{a^3/(GM)}$ must be a time (where a is some length) in our study of the free-fall problem. Here we get exactly the same thing, except for the factor of 2π.

We usually write Newton's version of Kepler's Third Law as

$$\boxed{P^2 = \frac{4\pi^2}{GM}a^3 = \frac{4\pi^2}{G(m_1 + m_2)}a^3} \tag{6.47}$$

An even more remarkable (and useful) thing about this relation is that it takes two quantities that are relatively easily accessible observationally (P and a) and relates them to one for which we would otherwise have a difficult time obtaining data, namely M, the total mass of the system.

Example: the mass of the Sun follows from two facts we already know:

$$P = 1 \text{ yr} = 3.16 \times 10^7 \text{ s} \tag{6.48}$$

$$a = 1 \text{ AU} = 1.5 \times 10^{11} \text{ m} \tag{6.49}$$

$$G = 6.67 \times 10^{-11} \frac{\text{m}^3}{\text{kg s}^2} \tag{6.50}$$

thus

$$M_\odot + M_{\text{earth}} = \frac{4\pi^2}{GP^2}a^3 = 2 \times 10^{30} \text{ kg} \tag{6.51}$$

We can also immediately see that Kepler's Third law isn't quite right, since it makes no mention of the masses of the planets (which are different from one another). However, since $M_\odot \gg M_{\text{planet}}$ for our solar system it is approximately true that

$$P^2 = \frac{4\pi^2}{GM_\odot}a^3 \tag{6.52}$$

and you can verify that if P is in years and a in AU, then

$$\frac{4\pi^2}{GM_\odot} = 1\frac{\text{yr}^2}{\text{AU}^3} \tag{6.53}$$

You should compare this to Equation 2.14. Historically, Newton actually went from Kepler's relation to the inverse square law for gravity! So this law is really a reflection of the specific form of the gravitational force, unlike Kepler's Second Law, which would be true for any force which is directed along the line between the two particles (a so-called "central force").

6.3 Exercise: Calculating Energy and Angular Momentum in the Two-body Problem

The goals for this exercise are
- Understand how to calculate the energy and angular momentum for the (fictitious) two-body problem
- Understand the relation between energy and semimajor axis of the orbit
- Understand how to create dimensionless quantities for numerical integration, and re-apply dimensions to extract physical quantities
- Understand how the conservation (time-independence) of energy and angular momentum comes about, even though they are composed of time-dependent quantities

We will use the following information about a two-body system of a planet going around a star:

```
# Masses of the objects
m1 = c.M_sun
m2 = c.M_earth
# Initial conditions of the fictitious problem
x0 = 1.*u.AU
vx0 = 0.*u.m/u.s
y0 = 0.*u.m
vy0 = 3.3e4*u.m/u.s
```

Question 1. Recall that the energy in the fictitious two-body problem can be written as in Equation (6.19)

$$E = \frac{1}{2}\mu \, |\dot{\mathbf{r}}|^2 - \frac{GM\mu}{|\mathbf{r}|} = \frac{1}{2}\mu v^2 - \frac{GM\mu}{r} \tag{6.54}$$

where

$$M = m_1 + m_2 \tag{6.55}$$

and

$$\mu = \frac{m_1 m_2}{m_1 + m_2} \tag{6.56}$$

We know that energy is conserved, which means if you calculate it at the initial time, it is the same ever after.

- Write two separate functions, one to calculate the kinetic energy and one to calculate the gravitational (this will come in handy later). Your functions should be able to handle multiple values of position or velocity.
- Calculate the total energy by using the values of position and velocity given in the initial conditions.
- What do you notice about the sign of the energy? What does this mean physically?

Question 2. We also know that we can calculate the semimajor axis of an orbit if we know the energy, using

$$a = \frac{GM\mu}{2|E|} \tag{6.57}$$

Write a `Python` function to calculate the semimajor axis a given m_1, m_2, and E, and calculate it for this problem.

Question 3. We would like to use our non-dimensional differential equation solver `TwoBody2DCart` to solve this problem. First, we need to convert the initial position and velocity to dimensionless form using a characteristic length, which we will take to be

$$r_c = a \tag{6.58}$$

the semimajor axis that you calculated above. (Note that I have moved away from notation like x_0 to denote the characteristic length to avoid confusing it with the initial condition. In the free-fall problem, we took the characteristic length to be the initial position, but here we are not.) Recall that the characteristic time is

$$t_c = \sqrt{\frac{r_c^3}{GM}}$$

(6.59)

Then a characteristic velocity is

$$v_c = \frac{r_c}{t_c}$$

(6.60)

and non-dimensional initial conditions are

$$X_0 = \frac{x_0}{r_c}$$

(6.61)

$$Y_0 = \frac{y_0}{r_c}$$

(6.62)

$$V_{X0} = \frac{v_{x0}}{v_c}$$

(6.63)

$$V_{Y0} = \frac{v_{y0}}{v_c}$$

(6.64)

Notice that these capital variables are dimensionless.

- What are the dimensionless initial conditions for this problem? What is the range of τ that corresponds to one period? Hint: use Newton's version of Kepler's Third Law and look at the definition of the characteristic time with $r_c = a$. (Note that the way we've set things up, the dimensional period is *not* t_c, and thus the non-dimensional period is not $\tau = 1$.)
- Use `solve_ivp` in `TwoBody2DCart` to solve for the orbit of the planet with these initial conditions over a range of τ corresponding to one period. Make a plot of this orbit X vs Y.
- Convert the output back to dimension-full quantities, and make another plot of the orbit (x versus y), labeling the axes. Can you convince yourself that the semimajor axis is as expected from Question 3?

6.4 Study Questions

Problems that are computational are indicated with a $\star$. In general, keeping three significant figures (two numbers after the decimal place in scientific notation) is sufficient: for example, $M_{earth} = 5.97 \times 10^{24}$ kg.

In both problems below, you may consider the Earth's orbit around the Sun to be perfectly circular with a radius (which is also the semimajor axis) of 1 AU (astronomical unit), and a period of 1 year (= 31557600 s). (I kept more than 3 significant figures here because the definition is exact: $365.25 \times 24 \times 3600$ s.)

You can also assume that there is negligible gravitational interaction between the Earth and asteroid in Question 1 and between the Earth and the comet in Question 2. That is, the orbital motions of all bodies are just determined by their interaction with the Sun. All orbits also occur in the same plane.

1. An asteroid of mass $m_{ast} = 7.00 \times 10^{10}$ kg is also traveling around the Sun with a period of 2.15 year and an eccentricity of 0.4. Astronomers predict

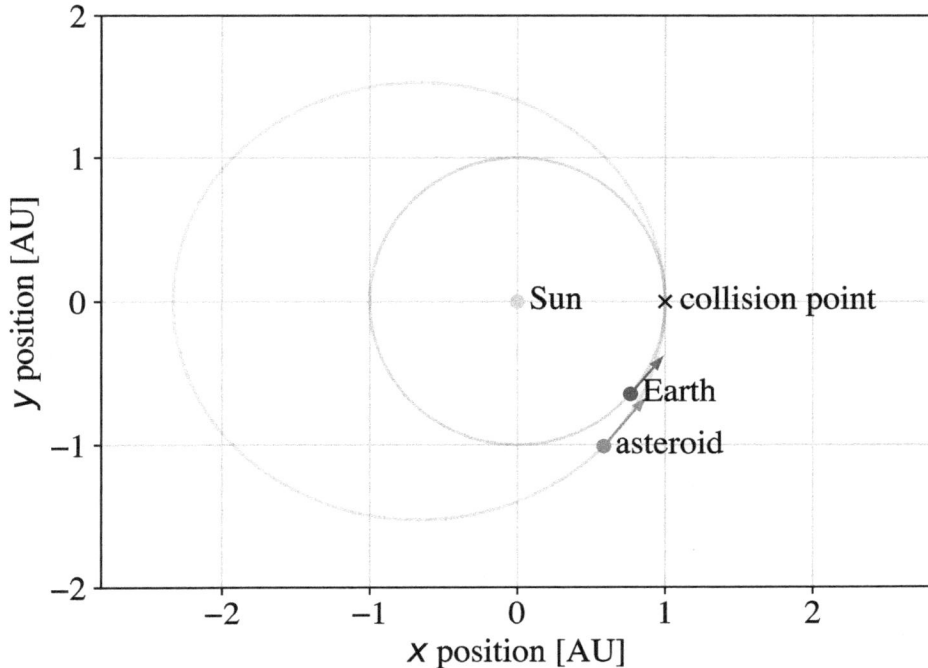

Figure 6.1. The orbits of the Earth and asteroid are shown in blue and red, respectively, with the sense of orbital velocity shown as arrows (these are not to scale). The Earth and asteroid are shown on their orbits before the collision; the collision point at perihelion of the asteroid's orbit is shown with an X.

that the asteroid and the Earth are going to collide in the orientation shown in Figure 6.1, when the asteroid is at perihelion (closest approach to the Sun). Both the asteroid and the Earth are moving counterclockwise (in the sense of the arrows shown).

(a) What is the semimajor axis of the asteroid's orbit?

(b) ⋆ Convert the semimajor axis to km, AU, and lightyears using `astropy` unit conversions.

(c) What is the total energy of the asteroid's orbit around the Sun? (in Joules) Compare this to the total energy of Earth's orbit around the Sun.[2]

(d) What is the angular momentum of the asteroid's orbit (in Joule-seconds)?

(e) What is the *vector* velocity of the Earth $\mathbf{v}_{Earth} = v_x \hat{\mathbf{x}} + v_y \hat{\mathbf{y}}$ in its orbit at the collision point, using the coordinate system shown in Figure 6.1? (Do not look this up; use the values in this problem to find the Earth's velocity.)

---

[2] In most cases in this book, I use "compare" to mean "tell many how many times larger or smaller one quantity is than another". That is, a *ratio*, not a difference. This is by far the more useful comparison when two quantities differ by orders of magnitude.

(f) Similarly, what is the vector velocity of the asteroid at the collision point (when it intercepts the Earth)? *Hint*: Look back at the Exercise at the end of Chapter 5 for the orientation of the velocity when an orbit is at perihelion and use Part 1d.

(g) Assuming that the gravitational attraction of the Earth does not accelerate the asteroid, and ignoring the finite size of the Earth, what is the relative vector velocity difference $\Delta \mathbf{v} = \mathbf{v}_{\text{Earth}} - \mathbf{v}_{\text{ast}}$ between the Earth and the asteroid just before the collision?

(h) What happens during the collision is fairly complicated. We will make some simplifying assumptions. As indicated, we are going to neglect the gravitational interaction between the Earth and the asteroid (not a great approximation). This means that, before the collision, the total energy of the system is given by the sum of the asteroid's and Earth's energies. After the collision, we assume the total mass of the Earth is $M_{\text{earth}} + m_{\text{ast}}$ (all of the asteroid's mass sticks around) and that neither the position nor the velocity of the Earth's orbit are changed (a pretty good approximation). Under these assumptions, show that the portion of the total energy due to the gravitational potential energy is the same before and after the collision by considering the situation immediately before and immediately after the collision.

(i) The kinetic energy of the asteroid will change, and the energy released by the impact in this case is approximated by $\frac{1}{2}m_{\text{ast}}|\Delta \mathbf{v}|^2$, where you found $\Delta \mathbf{v}$ in Part 1g. (Notice that if the asteroid is moving along at exactly our velocity, the effect of the impact is minimal; it really is the relative velocity that matters.) Acknowledging that this is a pretty dire scenario, compare this energy release from the change in kinetic energy to the energy released by the Hiroshima nuclear bomb, 53×10^{12} J.

2. Now consider a comet of mass $m_{\text{comet}} = 2.2 \times 10^{14}$ kg with a semimajor axis of 600 AU and an eccentricity of 0.999. The comet's orbit relative to Earth is shown in Figure 6.2. We assume the comet and the Earth start at $t = 0$ with the Earth at $x = 1$ AU, $y = 0$ and the comet at its perihelion as shown.

(a) Referring to Figure 6.2, at what two values θ_1 and θ_2 does the comet's orbit intersect the Earth's? *Hint*: Both the comet and the Earth have the Sun at one focus of their orbits.

(b) To set up a numerical two-body problem for the comet, we're going to need initial conditions. Determine these for the comet by determining the distance and velocity at perihelion (in physical units) similarly to Item 1f. Turn these into non-dimensional initial conditions using our usual trick of $r_c = a$ and $t_c = \sqrt{a^3/(GM)}$. Note that determining these numbers doesn't require a numerical solution, so I have not put a ⋆ on this problem.

(c) ⋆ Using the initial conditions from 2b, use our Python code to solve the initial value problem for the orbit, and make a plot of the comet's

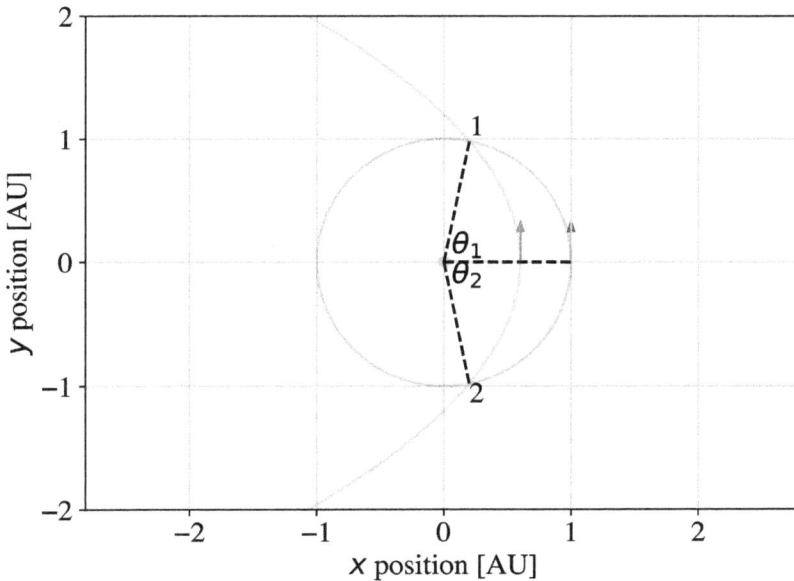

Figure 6.2. The orbits of the Earth and comet are shown in blue and red, respectively, with the sense of orbital velocity shown as arrows (these are not to scale). The two possible collision points are marked 1 and 2, located at the polar coordinates $r = 1$ AU and $\theta = \theta_1$ or $\theta = \theta_2$.

orbit from perihelion to a point well past the intersection Point 1 with the Earth's orbit. You will need to "play" with the range of τ necessary to get the solution to include just this part of the comet's orbit. Make the plot in physical units x, y (meters or AU).

(d) ⋆ Convert the output x- and y-coordinates from the numerical solution of the comet's orbit in 2c to r and θ and plot both r vs t and θ vs t. Find the time at which the comet reaches Point 1 by finding when the distance r reaches $r = 1$ AU and $\theta = \theta_1$. Do this graphically by plotting $r = 1$ AU on your $r(t)$ plot using `plt.axhline(R)` where R is 1 AU, and similarly for θ_1. You should also find the nearest time in the numerical solution for t by searching in the list of r and θ.

```
# Returns the index in the array r which
# is closest to the value R.  That is,
# the index where the argument of np.argmin is
# a minimum
wh = np.argmin(np.abs(r - R))
# This will return the value of t where
# r is closest to R
t_point1 = t[wh]
```

You may need to have `solve_ivp` increase the number of points in τ to get a satisfactory answer.

(e) How long does it take the Earth to reach the Point 1 (at θ_1?) (Don't overthink this; you don't need the numerical solution of an orbit for this part.)

(f) ⋆ Does the comet reach Point 1 before, after, or at the same time as the Earth? Explain your reasoning. (The numerical solution will be helpful, but other reasoning can be used as well.)

An Introduction to Astrophysics with Python
Stars and planets
James Aguirre

Chapter 7

Velocities in the Two-body Problem

Here we consider the time dependence of both position and velocity in the two-body problem, and since we must calculate the position numerically, we must also calculate the velocity numerically. We thus consider the complement of numerical integration, numerical differentiation. We use this to explore quantitatively how energy and angular momentum are constant in time, in spite of the position and velocity of the two bodies changing continuously.

7.1 The Vector Components of Velocity

The two-body problem has been incredibly well studied over its 400 year history, so that there is an analytic result for basically any question you might care to ask (Reed 2023; though some of the results are rather involved). Here we look at some examples of this. For the case of the elliptical orbit, it is fairly straightforward to work out the components of the velocity in polar coordinates, at least implicitly in terms of their dependence on θ. As always, the dependence on time requires some appeal to a numerical method, as we have seen.

We can start with the known equation for the elliptical orbit

$$r(\theta) = \frac{a(1 - e^2)}{1 + e \cos \theta} \tag{7.1}$$

and recall that the components of the velocity are

$$v_r = \frac{dr}{dt} = \frac{dr}{d\theta}\frac{d\theta}{dt} = \frac{dr}{d\theta}\dot{\theta} \tag{7.2}$$

$$v_\theta = r\frac{d\theta}{dt} = r\dot{\theta} \tag{7.3}$$

doi:10.1088/2514-3433/ade5f6ch7

Working out $dr/d\theta$ is straightforward:

$$\frac{dr}{d\theta} = \frac{a(1 - e^2)e \sin \theta}{(1 + e \cos \theta)^2} \tag{7.4}$$

$$= r\frac{e \sin \theta}{1 + e \cos \theta} \tag{7.5}$$

$$= r^2 \frac{e \sin \theta}{a(1 - e^2)} \tag{7.6}$$

for $\dot{\theta}$, let's make use of

$$\dot{\theta} = \frac{L}{\mu r^2} \tag{7.7}$$

and, by combining the eccentricity and semi-major axis equations, we get

$$L = \mu\sqrt{GMa(1 - e^2)} \tag{7.8}$$

which directly gives us

$$v_r = r^2 \frac{e \sin \theta}{a(1 - e^2)} \frac{\mu\sqrt{GMa(1 - e^2)}}{\mu r^2} \tag{7.9}$$

$$= \sqrt{GMa(1 - e^2)} \frac{e \sin \theta}{a(1 - e^2)} \tag{7.10}$$

$$= \sqrt{\frac{GM}{a(1 - e^2)}} e \sin \theta \tag{7.11}$$

Notice that $v_r = 0$ for a circular orbit $e = 0$, as it must be, and also that it vanishes at $\theta = 0$ and $\theta = \pi$ for any eccentricity. Similarly,

$$v_\theta = r\dot{\theta} \tag{7.12}$$

$$= \frac{1}{r}\sqrt{GMa(1 - e^2)} \tag{7.13}$$

$$= \sqrt{\frac{GM}{a(1 - e^2)}}(1 + e \cos \theta) \tag{7.14}$$

For a circular orbit with $r = a = R$ and $e = 0$, we recover the correct equation for the circular velocity.

We can also obtain an expression for the total velocity by noticing that

$$v^2 = v_r^2 + v_\theta^2 \tag{7.15}$$

$$= \frac{GM}{a(1-e^2)}((1 + e\cos\theta)^2 + e^2\sin^2\theta) \tag{7.16}$$

$$= \frac{GM}{a(1-e^2)}(1 + 2e\cos\theta + e^2\cos^2\theta + e^2\sin^2\theta) \tag{7.17}$$

$$= \frac{GM}{a(1-e^2)}(1 + 2e\cos\theta + e^2) \tag{7.18}$$

$$= \frac{GM}{a(1-e^2)}(2 + 2e\cos\theta + e^2 - 1) \tag{7.19}$$

$$= \frac{GM}{a(1-e^2)}(2(1 + e\cos\theta) - (1 - e^2)) \tag{7.20}$$

$$= GM\left(2\frac{1 + e\cos\theta}{a(1-e^2)} - \frac{1-e^2}{a(1-e^2)}\right) \tag{7.21}$$

$$= GM\left(\frac{2}{r} - \frac{1}{a}\right) \tag{7.22}$$

which is the so-called *vis-viva* equation. A clever re-write of this notices that it can also be written in terms of the period of the orbit

$$v^2 = \frac{4\pi^2 a^3}{P^2}\left(\frac{2}{r} - \frac{1}{a}\right) \tag{7.23}$$

with the obviously correct answer for a circular orbit.

Figure 7.1 shows the orientation of the vector velocity around an elliptical orbit. Note the magnitude of the vector follows from what expect based on Kepler's Second Law, namely that the speed is greatest at pericenter (the point of closest approach to the center of mass), and least at when it is furthest (apocenter). Note also that the velocity vector is tangent to the trajectory: this is true in general in mechanics, and is not specific to the two-body problem. In the two-body problem, though, it does mean that the velocity is purely tangential at the peri- and apocenter, but in general there will be both a radial and tangential component to the velocity at other points in the orbit.

7.2 Numerical Differentiation

We're often interested in derivatives of functions of one or more variables where we do not have an explicit, analytic form for the function to be evaluated. Mathematically, the functions of interest are usually continuous, but when we represent them numerically, we only have the function sampled at discrete points.

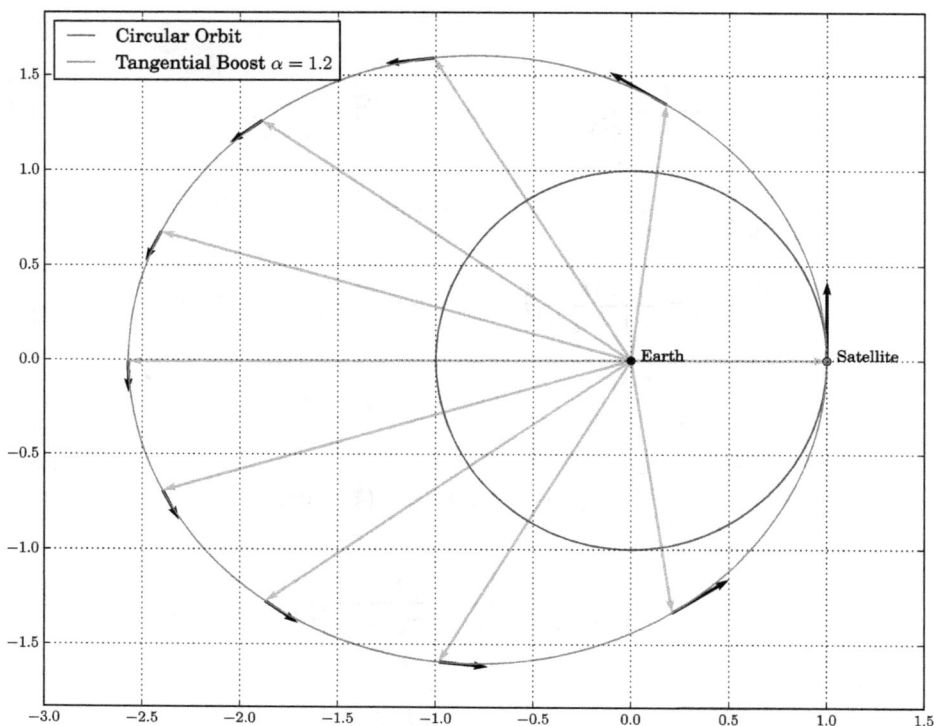

Figure 7.1. The vector velocity around an elliptical orbit. The position vector is shown in cyan and the velocity vector in black, tangent to the orbital trajectory in green.

We'd like to understand how we can, nevertheless, hope to approximate the derivative of such functions.

For example, suppose I have a list of times[1]

$$t = [t_1, t_2, t_3, ..., t_N] \tag{7.24}$$

Corresponding to this list t, we can associate values of a function $f(t)$ evaluated at those times, i.e.,

$$f = [f(t_1), f(t_2), f(t_3), ...f(t_N)] \equiv \left[f_1, f_2, f_3, ...f_N \right] \tag{7.25}$$

If we want to evaluate the derivative of f, we can simply go back to the definition of the derivative from calculus:

$$\frac{df}{dt} = \lim_{h \to 0} \frac{f(t+h) - f(t)}{h} \tag{7.26}$$

---

[1] Sometimes you will see this object called a "vector", but that word has a very specific meaning mathematically, and we're not always going to treat the object t as a mathematical vector, so I choose to just call it a "list".

Now, unless $h \to 0$ exactly, this is not the true mathematical derivative, but we might guess that if h is very small, then the resulting thing should be "close" to the true derivative, in some sense. For our discrete numerical function sampled at points t, which are very close together, we can then say

$$\frac{df}{dt}(t_1) \approx \frac{f(t_2) - f(t_1)}{t_2 - t_1} \equiv \frac{\Delta f}{\Delta t}(t_1) \tag{7.27}$$

Note that the value of the derivative at one point depends on the function value at that point *and* at the next timestep. More generally, for a point t_i with $1 \leqslant i \leqslant N$

$$\frac{df}{dt}(t_i) \approx \frac{f(t_{i+1}) - f(t_i)}{t_{i+1} - t_i} \tag{7.28}$$

You may notice that this will cause a small problem when $i = N$, since t_{N+1} and $f(t_{N+1})$ weren't in my original lists. If I can generate those points somehow, then I can evaluate the last point, but if not, the simplest thing is to just have one less point in the list defining the derivative than you did in the original function.

7.3 The `Python` Implementation

The `Python` function np.diff lets us generate this numerical approximation fairly easily. If I have a list t

$$t = [t_1, t_2, t_3, \ldots, t_N] \tag{7.29}$$

then

$$\mathtt{np.diff} = [t_2 - t_1, t_3 - t_2, \ldots, t_N - t_{N-1}] \tag{7.30}$$

Thus our numerical approximation can be written in `Python` as

$$\mathtt{diff(f)/diff(t)} = \left[f_2 - f_1, f_3 - f_2, \ldots f_N - f_{N-1} \right] / \tag{7.31}$$
$$[t_2 - t_1, t_3 - t_2, \ldots, t_N - t_{N-1}]$$

Note that the list produced by np.diff has $N - 1$ elements.

Now let's look at the statement that this numerical approximation is "close" to the true derivative. How close is it? For that, recall the Taylor series expansion of a function

$$f(x + h) = f(x) + hf'(x) + \frac{h^2}{2!}f''(x) + \frac{h^3}{3!}f'''(x)\ldots \tag{7.32}$$

This statement is exact (if we include infinitely many terms). We can rearrange this to get

$$f'(x) = \frac{f(x + h) - f(x)}{h} - \frac{h}{2!}f''(x) + \cdots \tag{7.33}$$

The first term is just our approximation for the derivative. So the discrepancy between our approximation and the true derivative is

$$-\frac{h}{2!}f''(x) \tag{7.34}$$

plus terms that go to zero as powers of h (i.e., faster than h if $h \ll 1$).

A simple way to do better is to notice that

$$f(x - h) = f(x) - hf'(x) + \frac{h^2}{2!}f''(x) + \cdots \tag{7.35}$$

so that if we take the difference between Equations (7.32) and (7.35) we get

$$f'(x) = \frac{f(x + h) - f(x - h)}{2h} - \frac{h^2}{3!}f'''(x) + \cdots \tag{7.36}$$

so this approximation has errors that go to zero as h^2 (times the third derivative), instead of as h. It's also more symmetrical, with the estimate of the derivative at a point depending on the previous point and the next point. (This will also create a problem for the first estimate in the list, since it won't have a previous point.)

There are increasingly sophisticated formulas that will decrease the error in the numerical approximation for a fixed value of timestep h, and also formulas that deal better with cases in which f has some unusual behavior (like becoming singular) but we will not be concerned with them here.

We finally note that `numpy` has a built-in function to compute numerical derivatives, called `np.gradient`. You'll recall that the mathematical definition of the gradient of a function f(x,y,z) is

$$\nabla f = \frac{\partial f}{\partial x}\hat{\mathbf{x}} + \frac{\partial f}{\partial y}\hat{\mathbf{y}} + \frac{\partial f}{\partial z}\hat{\mathbf{z}} \tag{7.37}$$

In the case that f is only a function of x, the gradient reduces to our usual definition of derivative, so

$$\texttt{gradient(f, x)} = \frac{df}{dx} \tag{7.38}$$

`np.gradient` has the nice feature that it uses Equation (7.32) to calculate the derivative at the first data point, but (7.35) for the data point and Equation (7.36) in the middle, so the resulting array has the same size as the input, and the higher precision of Equation (7.36) for most values. Examples of the use the use of `Pygradient` to compute numerical derivatives of a function, and the effect on the accuracy depending on the step size h, are shown in Figures (7.2 and 7.3)

7.4 Exercise: Two-body E and L Time Dependence

The goals of this exercise are:
- Explore the time dependence of E and L in the two-body problem by explicitly calculating them
- Become aware of the limitations of the numerical calculation
- Learn how to calculate derivatives numerically
- Gain a better understanding of the velocity of the two-body orbit expressed in polar coordinates

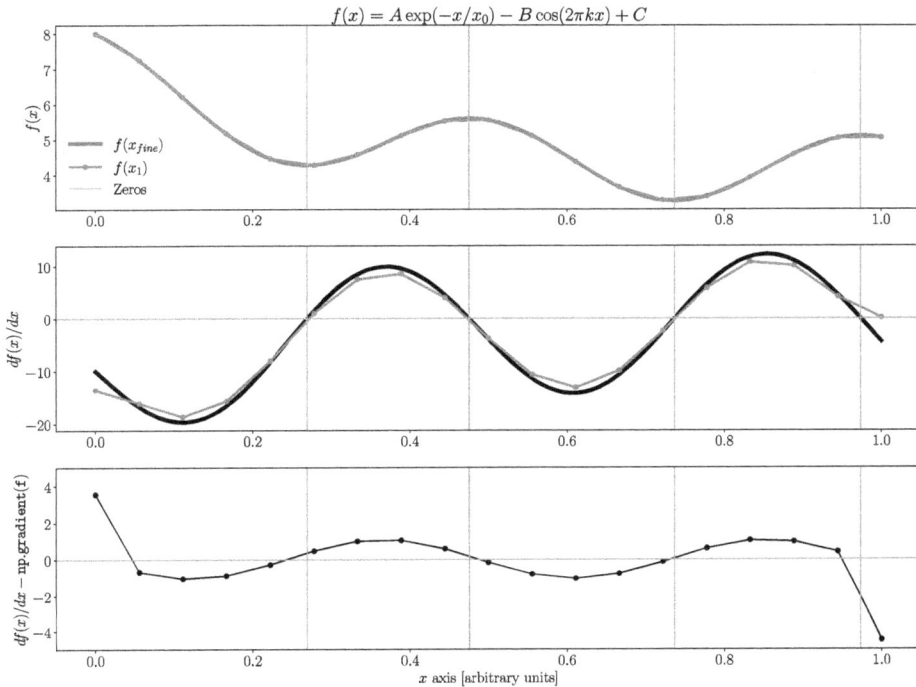

Figure 7.2. An example of taking numerical derivatives using the `np.gradient` function with a relatively coarse gridding in the dependent variable. In this case, we have an analytic function, so we can compute its derivative explicitly and compare to the numerical approximation. The top panel shows the function and the points at which it was sampled. The middle panel shows the derivative and the `np.gradient` approximation to it, and the bottom panel shows the difference between the true derivative and the numerical approximation. Note that the numerical version has large errors at the endpoints (due to only being able to estimate the derivative to one side). We are also interested in the zeroes of the derivative (where it reaches an extremum) and clearly the numerical approximation also does a poor job of locating these zeroes (whose positions are indicated with vertical lines).

We will be using the same setup as the Exercise in Chapter 6.

Question 1. Use the functions defined above to separately calculate the gravitational potential energy and kinetic energy of the planet as a function of time. The SI units of energy are Joules (J), where $1 \text{ J} = 1 \text{ N m} = 1 \text{ kg m}^2 \text{ s}^{-2}$. Plot both the kinetic and potential energies versus time on the same plot, together with their sum (the total energy). Label the axes! Use a legend to identify the various curves. Explain how the total energy *should* behave as a function of time and comment on whether or not your plot agrees with these expectations. Use `plt.axhline(0, color='gray')` plot a horizontal line across the plot) to show where 0 energy is.

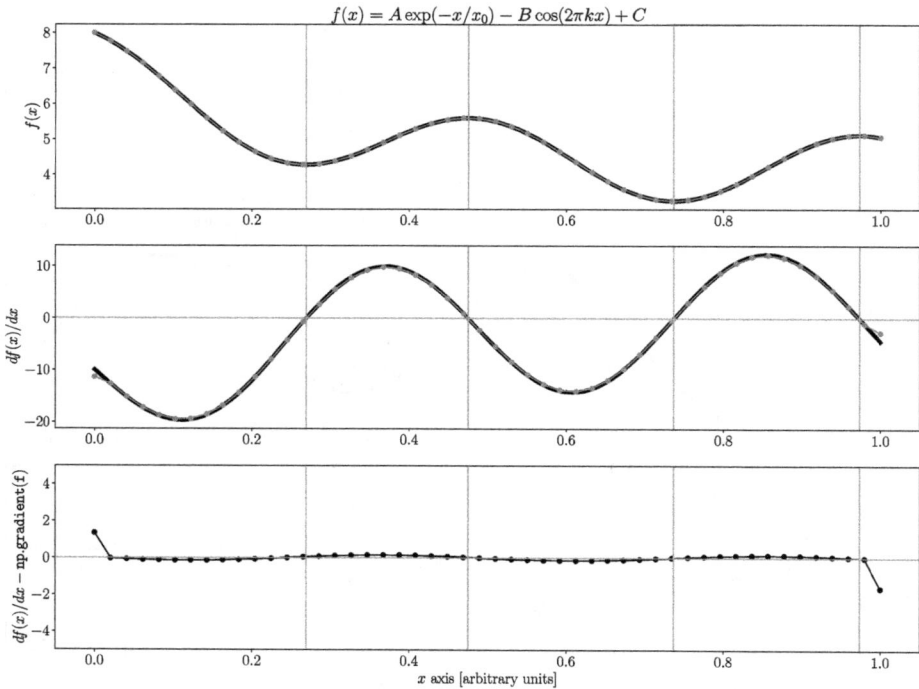

Figure 7.3. The same as Figure 7.2, but with a finer gridding (the function is evaluated at points closer together). The errors are smaller overall (but still relatively large at the endpoints), but the estimate of the zeroes is clearly improved.

We can calculate $r(t)$ and $\theta(t)$ from $x(t)$ and $y(t)$ using the relations

$$r(t) = \sqrt{x(t)^2 + y(t)^2} \tag{7.39}$$

$$\theta(t) = \arctan(y/x) \tag{7.40}$$

The arctan is tricky (why?); you'll find numpy's arctan2 function helpful. It works like this:

```
theta = np.arctan2(y,x)
```

and returns values $-\pi \leqslant \theta \leqslant \pi$. We generally like $0 \leqslant \theta \leqslant 2\pi$, and numpy anticipates that we will often want to put angles into this range. We can use the function 'np.unwrap as follows:

```
theta = np.unwrap(np.arctan2(y,x))
```

Question 2. Calculate $r(t)$ and $\theta(t)$ from $x(t)$ and $y(t)$ by writing a function `cart2pol`. (Recall that in Exercise I.5 we wrote a function `pol2cart` to calculate polar coordinates from Cartesian ones.) Plot r and θ versus time on a multi-panel plot. Also, plot r, θ on a polar plot ('plt.polar') to show the orbit. Make sure you get a plot for the orbit that has the same semi-major axis as we saw in the Chapter 6 exercise.

Calculating derivatives numerically can be done with the numpy function `np.gradient`. The syntax for computing the (one-dimensional) derivative df/dt of a function $f(t)$ represented by values `f` at times `t` is

```
dfdt = np.gradient(f, t)
```

Question 3. The vector velocity can be written as

$$\mathbf{v} = v_r \hat{r} + v_\theta \hat{\theta} \tag{7.41}$$

Compute $v_r = \dot{r}$ and $v_\theta = r\dot{\theta}$ numerically. Plot $v_r(t)$ and $v_\theta(t)$ versus time on the same plot. Which component is in general larger? At which points (if any) is the velocity purely in the $\hat{\theta}$-direction? At which points (if any) is the velocity purely in the $\hat{r}$-direction? It may help to add a horizontal line to show where 0 velocity is, similar to what you did for zero energy in Question 1.

7.5 Study Questions

1. Recall how we non-dimensionalized the two-body problem. We picked a characteristic length r_c (a number in meters, to be specified later), and rewrote all of the lengths in the problem as, for example, $x = r_c X$ where X is a dimensionless number. We then noticed that, if we also picked a characteristic time

$$t_c = \sqrt{\frac{r_c^3}{GM}} \tag{7.42}$$

then the equation of motion for the fictitious problem could be written in terms of a dimensionless position vector $\mathbf{X}$ as a function of a dimensionless time $\tau = t/t_c$

$$\mathbf{X} = \begin{bmatrix} X(\tau) \\ Y(\tau) \end{bmatrix} = \begin{bmatrix} \dfrac{x(t/t_c)}{r_c} \\ \dfrac{y(t/t_c)}{r_c} \end{bmatrix} \tag{7.43}$$

with equation of motion

$$\frac{d^2\mathbf{X}}{d\tau^2} = -\frac{1}{|\mathbf{X}|^2}\hat{\mathbf{X}} \tag{7.44}$$

(This isn't *exactly* the way I wrote it in Chapter 5, so make sure you're happy with the notation before proceeding.) We can see what happens in polar coordinates by remembering the relation between polar r and Cartesian:

$$r = \sqrt{x^2 + y^2} \tag{7.45}$$

Let's make this dimensionless by dividing both sides by r_c:

$$\frac{r}{r_c} = \frac{1}{r_c}\sqrt{x^2 + y^2} = \sqrt{\frac{x^2}{r_c^2} + \frac{y^2}{r_c^2}} = \sqrt{X^2 + Y^2} \tag{7.46}$$

and we define a new dimensionless polar coordinate

$$\rho \equiv |\mathbf{X}| = \sqrt{X^2 + Y^2} \tag{7.47}$$

with $r = r_c\rho$.

Remember that derivatives end up looking like, for example,

$$\frac{dr}{dt} = \frac{d(r_c\rho)}{d(t_c\tau)} = \frac{r_c}{t_c}\frac{d\rho}{d\tau} \tag{7.48}$$

In the exercises, we also defined a "characteristic velocity"

$$v_c = \frac{r_c}{t_c} \tag{7.49}$$

Now that we know how important energy and angular momentum are in the two-body problem, let's also define a characteristic energy and angular momentum

$$E_c = \mu\frac{r_c^2}{t_c^2} \tag{7.50}$$

$$L_c = \mu r_c v_c = \mu\frac{r_c^2}{t_c} \tag{7.51}$$

where we have just used dimensional analysis to combine mass (specifically, the reduced mass μ), length, and time to make these quantities. The dimensionless energy and angular momentum are then

$$\varepsilon = \frac{E}{E_c} \tag{7.52}$$

$$\lambda = \frac{L}{L_c} \tag{7.53}$$

(a) Write the two-body energy

$$E = \frac{1}{2}\mu\, |\dot{\mathbf{r}}|^2 - \frac{GM\mu}{r} \tag{7.54}$$

in dimensionless form. (Hint: the result will have ε expressed only as a function of ρ, its derivatives, and pure numbers).

(b) If a is the semi-major axis of an elliptical orbit, express the ratio a/r_c in terms of the dimensionless energy ε.

(c) Express the eccentricity e of an elliptical orbit in terms of the dimensionless energy ε and dimensionless angular momentum λ.

(d) If we pick $r_c = a$, what is the relation between L_c and the angular momentum of a circular orbit of radius a?

(e) If we know for an orbit that $\varepsilon < 0$ (that is, it is bound), and we know that $e > 0$, is the angular momentum the same, greater, or less than a circular orbit with the same ε? Explain your reasoning, using the relations you derived above.

Reference

Reed, B. C. 2023, Keplerian Ellipses 2nd edn (Bristol: IOP Publishing)

AAS | IOP Astronomy

An Introduction to Astrophysics with Python
Stars and planets
James Aguirre

Chapter 8

Features of Orbits in the Two-body Problem

We now consider how we can classify different kinds of orbits based on their energy and angular momentum, considering both bound and unbound obits. We also consider the important special case of circular orbits. We consider in some detail how the initial conditions of the two-body problem determine the resulting orbit.

8.1 The Effective Potential

Let's go back to the expression Equation (6.24) for the energy of the two-body problem, written in terms *only* of the scalar function $r(t)$ (and its derivative), which gives the separation between the two bodies (given their total mass M, the reduced mass μ, and the angular momentum of the orbit L):

$$E = \frac{1}{2}\mu\dot{r}^2 + \frac{L^2}{2\mu r^2} - \frac{GM\mu}{r} \qquad (8.1)$$

The advantage of formulating the problem in this way is that the kinds of motions that a particle can have are determined by the function $r(t)$. Specifically, if $r(t) < R$ for some fixed value of R *for all times*, then we say the motion is *bound*. This means that, whatever else happens, the separation between the two objects cannot get arbitrarily large; they're always "stuck" together in some sense. But, if r can approach arbitrarily close to infinity, then the motion is *unbound*. The two possible scenarios are determined by the value of the energy E. Let us see how this comes about.

First we note that this energy equation is equivalent to a particle of mass μ moving in a *one-dimensional* "effective" potential

$$V_{\text{eff}} = \frac{L^2}{2\mu r^2} - \frac{GM\mu}{r} \qquad (8.2)$$

doi:10.1088/2514-3433/ade5f6ch8

The effective potential is a quantity with units of energy but which is *not* a potential energy that's derived from some simple force law. (If you did compute the gradient to find the "force" associated with this potential, you would find both a $1/r^2$ term from the gravitational potential, but also a positive $1/r^3$ associated, effectively, with the orbital kinetic energy.) Nevertheless, this formulation lets us ask how the particle moves in this fictitious mathematical one-dimensional problem. We're now going to divide this into a "kinetic" and "potential" energy purely by analogy with what we would write down for a 1D problem; obviously the real physics is 3D. Writing it this way gives the total energy as

$$E = K_r + V_{\text{eff}} \tag{8.3}$$

where

$$K_r = \frac{1}{2}\mu\dot{r}^2 \tag{8.4}$$

Note that this K_r is *not* the full kinetic energy, but only the term related to $\dot{r}$. The angular portion was absorbed into V_{eff}. We can make a plot of the various terms in the effective potential as shown in Figure 8.1.

Notice that V_{eff} has a minimum, which we will denote V_{min}, occurring at some value r_{min}. (We will determine r_{min} below.) The total energy $E = K_r + V_{\text{eff}}$ determines the allowed range for r: $r(t)$ can take on any such that $V_{\text{eff}} < E$, with the difference between V_{eff} and E made up by the "kinetic" term K_r. Since

$$K_r = \frac{1}{2}\mu\dot{r}^2 \geqslant 0 \tag{8.5}$$

it follows that $E \geqslant V_{\text{eff}}$, regardless of the sign of V_{eff}.

There are three cases of interest for the value of E:

1. $E = V_{\text{min}}$: This corresponds to K_r having its smallest value $K_r = \frac{1}{2}\mu\dot{r}^2 = 0$, which implies $\dot{r} = 0$, or $r = r_{\text{circ}} = $ const. This is thus a circular orbit.
2. $V_{\text{min}} < E < 0$: Then r must range only between some $r_{\text{min,ellip}}$ and $r_{\text{max,ellip}}$, corresponding to perihelion and aphelion.
3. $E \geqslant 0$: Then any $r > r_{\text{min}}$ is permitted, and the orbit is unbound. The orbit has a point of closest approach $r_{\text{min,para}}$ for $E = 0$, and $r_{\text{min,hyper}}$ for $E > 0$.

It turns out that *all* of these orbits are the so-called *conic sections*, the curves one gets when slicing a cone with a plane at various angles to the base of the cone. The circle and ellipse are the bound orbits, and the parabola and hyperbola are unbound. The parabola corresponds to the special case $E = 0$. This corresponds to an interesting condition, namely

$$\frac{1}{2}\mu v^2 - \frac{GM\mu}{r} = 0 \tag{8.6}$$

which implies a relation between the velocity necessary to achieve the parabolic orbit and the instantaneous distance between the two masses, namely

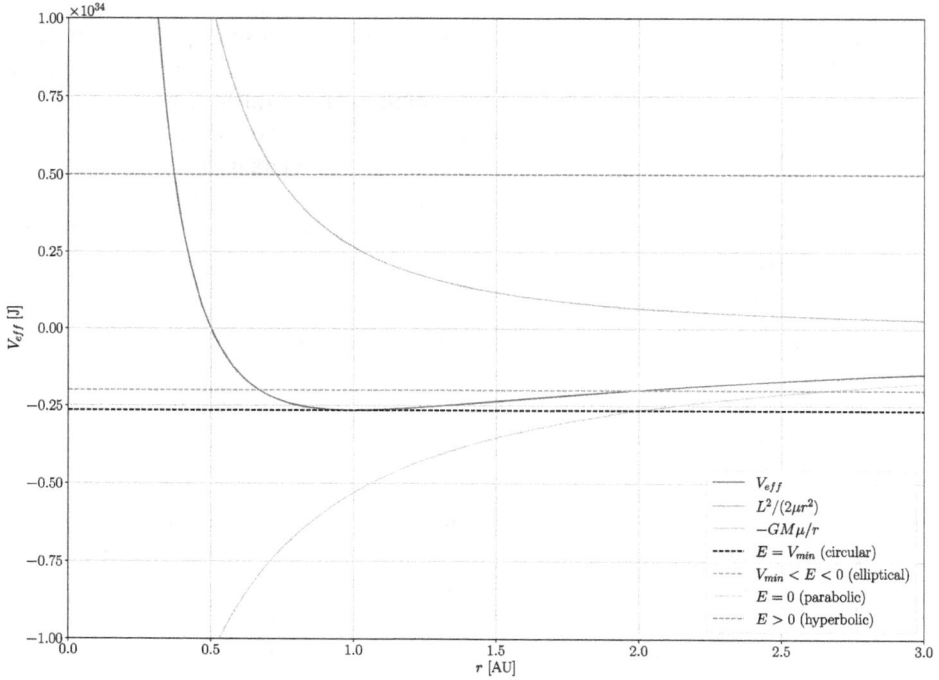

Figure 8.1. The various terms in the effective potential of the fictitious one-body problem. The numerical values are taken for the Earth–Sun system. Note that the permitted values of $r(t)$ are given by the portion of the curve lying *below* the marked energy. Notice that the range of energy between the circular orbit at the minimum of the effective potential and the unbound orbit at $V_{\text{eff}} = 0$.

$$v_{\text{esc}} = \sqrt{\frac{2GM}{r}} \qquad (8.7)$$

where v_{esc} is the so-called *escape velocity*. Thus, at a distance r, if the velocity of the orbit equals or exceeds v_{esc}, then the orbit is unbound, or if the velocity is changed to exceed v_{esc}, then that will make the orbit unbound.

8.2 The Special Case of Circular Orbits

Let's look a bit more at the special case of circular orbits. For circular orbits we have seen that E must be a minimum, $E = V_{\text{min}}$. Starting from

$$V_{\text{eff}} = \frac{L^2}{2\mu r^2} - \frac{GM\mu}{r} \qquad (8.8)$$

We can find where the radius r_{circ} at which the minimum occurs by finding the value r_{circ} where the derivative is zero:

$$\frac{dV_{\text{eff}}}{dr} = 0 = \frac{-L^2}{\mu r^3} + \frac{GM\mu}{r^2} \qquad (8.9)$$

and the value r_{circ} that solves this is

$$\boxed{r_{\text{circ}} = \frac{L^2}{GM\mu^2}} \tag{8.10}$$

Thus, the distance from the center at which the circular orbit occurs is set by the angular momentum. Plugging back in, we find

$$V_{\min} = \frac{L^2}{2\mu}\left(\frac{GM\mu^2}{L^2}\right)^2 - \frac{G^2M^2\mu^3}{L^2} \tag{8.11}$$

or

$$V_{\min} = -\frac{G^2M^2\mu^3}{2L^2} = -\frac{1}{2}\mu\left(\frac{GM\mu}{L}\right)^2 \tag{8.12}$$

Notice that by dimensional analysis, since $V_{\min}$ is an energy, the term in parenthesis $GM\mu/L$ must be a velocity.

Let's work out what the circular velocity would be at any distance r by noting that the magnitude of the angular momentum for the circular orbit is

$$L_{\text{circ}} = \mu r_{\text{circ}} v_{\text{circ}} \tag{8.13}$$

and plugging back into Equation (8.9)

$$-\frac{\mu^2 r_{\text{circ}}^2 v_{\text{circ}}^2}{\mu r_{\text{circ}}^3} + \frac{GM\mu}{r_{\text{circ}}^2} = 0 \tag{8.14}$$

so that

$$\boxed{v_{\text{circ}} = \sqrt{\frac{GM}{r_{\text{circ}}}}} \tag{8.15}$$

which gives the velocity the planet must have in order to maintain a circular orbit. Note that this velocity is only $\sqrt{2}$ less than the escape velocity, Equation (8.7). Also note that an explicit expression for the angular momentum for a circular orbit is

$$\boxed{L_{\text{circ}} = \mu r_{\text{circ}} v_{\text{circ}} = \mu r_{\text{circ}}\sqrt{\frac{GM}{r_{\text{circ}}}} = \mu\sqrt{GMr_{\text{circ}}}.} \tag{8.16}$$

Also, a fun fact is that if we plug the relation for r_{circ} back into the equation for v_{circ}, we get yet another relation for the circular velocity

$$v_{\text{circ}} = \sqrt{\frac{GM}{r_{\text{circ}}}} = \sqrt{\frac{G^2M^2\mu^2}{L^2}} = \frac{GM\mu}{L} \tag{8.17}$$

which is the *same* velocity that showed up in our expression for the value of $V_{\min}$ in Equation (8.12). We can also write the energy of a circular orbit more simply by substituting Equation (8.10) into Equation (8.12) to get

$$\boxed{E_{\text{circ}} = V_{\min} = -\frac{1}{2}\frac{GM\mu}{r_{\text{circ}}}} \tag{8.18}$$

All of these interconnected expressions might lead to a sense of vertigo, in that it's not entirely clear what determines what for circular orbits. The key thing to keep in mind is that circular orbits are quite special, and instead of having to specify *both E* and *L*, the specification of the fact that the orbit is circular and one other parameter locks everything else in (given M and μ). Let's take two examples.

1. Since $\dot{r} = 0$ for circular orbits, the radius of the orbit r_{circ} is constant. Specifying r_{circ} then specifies the orbital velocity v_{circ} (Equation (8.15)), the angular momentum L_{circ} (Equation (8.16)) and the energy, Equation (8.18).
2. As we will see, L_{circ} is special for circular orbits: it's the largest the angular momentum can be for a given semi-major axis a. Specifying L_{circ} then determines r_{circ} via Equation (8.10), which again specifies all the other parameters (v_{circ}, E_{circ}).

8.3 General Properties of Bound Orbits

A key feature of the bound two-body problem is that *everything essential about the orbit is specified just by knowing E and L*.[1] Recall that the semi-major axis depends only on the energy

$$a = \frac{GM\mu}{2|E|} \tag{8.19}$$

and the eccentricity depends on both the energy and angular momentum

$$e^2 = 1 - \frac{2|E|L^2}{G^2M^2\mu^3} \tag{8.20}$$

from which we can work out that the semi-minor axis also depends on E and L:

$$b = \sqrt{\frac{L^2}{2\mu|E|}} \tag{8.21}$$

Let's consider some implications of Equations (8.19) and (8.20). Suppose we place a mass m_2 at a distance R from m_1 ($m_1 \gg m_2$).

1. If we give m_2 a velocity $v_{\text{circ}} = \sqrt{GM/R}$ (Equation (8.15)) in the direction perpendicular to m_1, then it will go in a circular orbit.
2. Now suppose we give it the same velocity v_{circ}, but in a *different* direction at an angle θ relative to the line between the masses; see Figure 8.2. Then it has a different angular momentum

$$L = \mu v_{\text{circ}} R \sin \theta < \mu v_{\text{circ}} R = L_{\text{circ}} \tag{8.22}$$

---

[1] I say "essential" because we are free to pick the orientation of the coordinate system relative to the semi-major axis of the ellipse, but this doesn't change the shape of the orbit or the rate of motion along it.

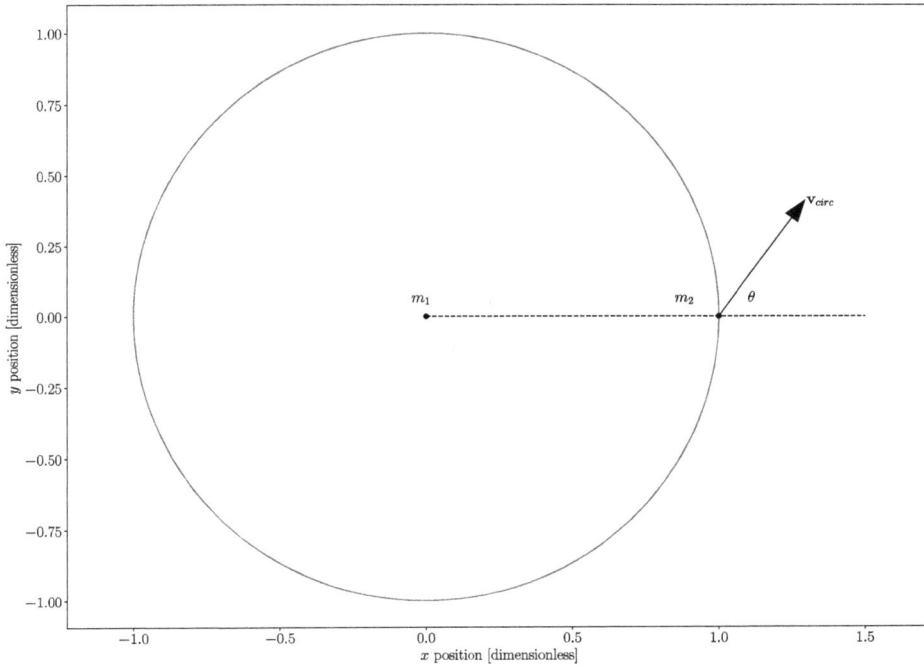

Figure 8.2. The orientation of a velocity at an angle θ relative to the line connecting m_1 and m_2. The circular orbit is shown, but the resulting orbit with the velocity shown at this position will be elliptical, with angular momentum *less* that the corresponding circular orbit, though the orbit will have the same semi-major axis.

and its orbit will not be circular, but elliptical, and the orbit will have $e > 0$ and an angular momentum *less* than it would have had on a circular orbit. However, since the energy is the same, it will have the same semi-major axis as the circular orbit. The resulting orbit is shown in Figure 8.3.

3. Consider another case in which we take the same masses, and start m_2 at different velocities, but always perpendicular to the $m_1 - m_2$ line. For $v = v_{\text{circ}}$ we of course get the circular orbit, and for $v < v_{\text{circ}}$, the orbit will be elliptical. (If $v = 0$, then of course m_2 falls directly into m_1 with eccentricity $e = 1$.) What happens if $v > v_{\text{circ}}$? Now we have to think a little. If $v_{\text{circ}} < v < v_e$, where v_e is the escape velocity, then the orbit will be bound (and elliptical). For $v = v_e$ we get a parabolic orbit, and for $v > v_e$ a hyperbolic orbit, both of which are unbound. Look back at Exercise I.5, Question 1, to explicitly see this happening as the velocity is increased.

4. There are two ways to get an orbit with $e = 1$: $|E| = 0$, corresponding to a parabola, and $L = 0$, corresponding to no tangential component to the velocity. For any other arrangement of $E < 0$, the eccentricity will be $e < 1$.

5. To make the eccentricity zero, I need to solve

$$0 = 1 - \frac{2|E|L^2}{G^2 M^2 \mu^3} \tag{8.23}$$

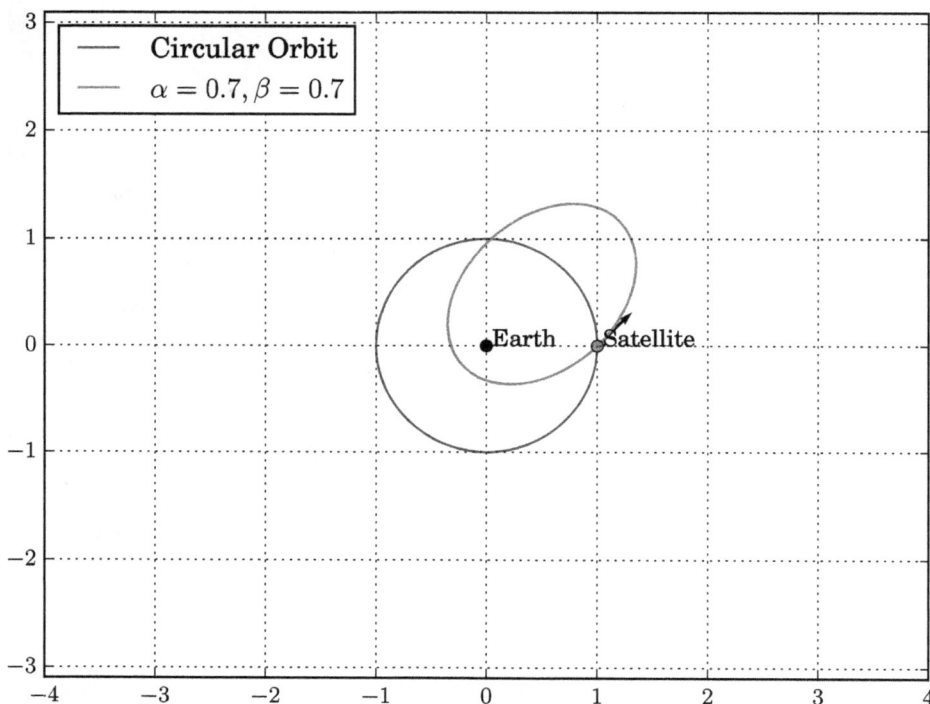

Figure 8.3. The orbit resulting from a changed velocity orientation. Both the original circular orbit and the new elliptical orbit are shown.

Substituting in the expression for the energy in terms of the semi-major axis, this gives

$$\frac{GM\mu L^2}{aG^2M^2\mu^3} = 1 \tag{8.24}$$

or

$$L_{\text{circ}} = \mu\sqrt{GMa} \tag{8.25}$$

is the special angular momentum required for a circular orbit of radius (semi-major axis) a, which is the same as we found in Equation (8.16) with $a = r_{\text{circ}}$. Notice that once we have specified $E < 0$ for a bound orbit, it must be true that any orbit with that energy has $L < L_{\text{circ}}$, or else the eccentricity would be imaginary. Thus, at fixed E the circular orbit has the *largest* angular momentum for an orbit with that semi-major axis.

8.4 Exercise: Features of Orbits

The setup here is the same as the last two exercises, namely the two-body problem with a star the mass of the Sun and a planet the mass of the Earth.

Question 1. At $t = 0$, compare the magnitude of the velocity of the planet in its orbit to the escape velocity. Write a `Python` function to calculate the escape velocity. Make sure the function returns a value in the correct units. Explain whether your answer is consistent with a bound orbit or an unbound orbit.

Question 2. How much energy would have to be added to the orbit to *just barely* make it unbound? Assume that we add this as kinetic energy at $t = 0$ with $x_0 = 1\,\text{AU}$ (the same initial position above). Take this additional energy, add it to the existing kinetic energy, and compute the new initial velocity for the unbound orbit using

$$E_{\text{kin}} + E_{\text{add}} = \frac{1}{2}\mu v_{\text{new}}^2 \tag{8.26}$$

Assume this velocity is again purely in the $\hat{y}$-direction. What would the new initial velocity be? Compare this with the escape velocity of Question 1.

What is the new *dimensionless* initial velocity?

Input the new dimensionless initial velocity (along with the original initial position and $V_{X0} = 0$) to the differential equation solver and show that you indeed get an orbit that looks parabolic rather than elliptical.

When constructing dimensionless quantities, it's useful to know there is a "dimensionless unit": `u.dimensionless_unscaled`. Above, we defined the initial position x0 in AU but the characteristic length r_c in meters. Note what happens:

```
print('Initial position x0:', Qprint(x0))
print('Characteristic length r_c:', Qprint(r_c))
print('Dimensionless initial position:', X0)

Initial position x0: 1 AU
Characteristic length r_c: 1.94e+11 m
Dimensionless initial position: 5.163487353316121e-12 AU /
    m
```

We can force a quantity with a ratio of units that are reduceable to a dimensionless quantity explicitly, and we can also extract just the numerical part or just the unit part as below. Notice that, just as we force our functions to return values in explicitly the correct unit, we want to get in the habit of converting quantities we know to be dimensionless to actually be so, or we run the risk of getting the wrong numerical value.

```
print('Numerical value of X0, not converting to
    dimensionless:', Qprint(X0.value))
print('Numerical value of X0, converting to dimensionless:'
    , Qprint(X0.to(u.dimensionless_unscaled).value))
print('Unit of X0, not converting to dimensionless:', X0.
    unit)
print('Unit of X0, converting to dimensionless:', X0.to(u.
    dimensionless_unscaled).unit)

Numerical value of X0, not converting to dimensionless:
    5.16e-12
Numerical value of X0, converting to dimensionless: 0.772
Unit of X0, not converting to dimensionless: AU / m
Unit of X0, converting to dimensionless:
```

So what we really should have done above was

```
X0 = (x0/r_c).to(u.dimensionless_unscaled).value
Y0 = (y0/r_c).to(u.dimensionless_unscaled).value
VX0 = (vx0/v_c).to(u.dimensionless_unscaled).value
VY0 = (vy0/v_c).to(u.dimensionless_unscaled).value
```

Question 3. Write a Python function to calculate the eccentricity e by using the energy E and the angular momentum L. (You needed to calculate the energy in Question 1, and here you will need to calculate the angular momentum; look back at the last exercise.)

What is e for this orbit? Verify that your function gives a dimensionless quantity by printing eccentricity.unit (assuming your variable is called eccentricity).

Question 4. Given a (again, go back and look at the exercise from last time) and e, what should $x(t = 0)$ be (in meters)? Recall what is special about the position on the ellipse which we've arranged for $t = 0$. Write your equation for $x(t = 0)$ in terms of a and e explicitly below. Verify that your initial position matches with both the dimensional and non-dimensional values used in calculating the orbit.

Question 5. Calculate the period P of this orbit (in seconds and in years). Another function might be useful here! What is the ratio of the period to the characteristic time with $r_c = a$

$$t_c = \sqrt{\frac{a^3}{GM_{\text{tot}}}} \qquad (8.27)$$

8.5 Study Questions

1. The escape velocity for the two-body problem is given by

$$v_{\text{esc}} = \sqrt{\frac{2GM}{R}} \qquad (8.28)$$

where R is the distance between the objects at a given time, and M is the sum of their masses. The escape velocity is not constant with time for a system (unlike energy or angular momentum). Also, when dealing with multiple bodies, it is often approximately correct to treat them as separate two-body systems, but one needs to think about which mass and distance are appropriate. This problem is intended to illuminate this issue.

One additional fact we will need in this problem is that spherical objects like the Earth and Sun can be treated, for the purposes of calculating a gravitational force, as if all of their mass were at a point at their center, and non-spherical objects that are small compared to the center-to-center distance between them can also be treated as points.

Because you are going to repeatedly evaluate the gravitational force between objects and the escape velocity, having a `Python` function handy to do this will make it much easier!

(a) Suppose we have a satellite of mass $m = 350$ kg sitting on the surface of the Earth. What is the gravitational force between the satellite and the Sun?[2] Between the satellite and the Earth? Based on this, if we want to know how fast a satellite needs to be moving to get far from the surface of the Earth (that is, to "unbind" itself from the Earth) which body (the Sun or the Earth) exerts the greater force, and therefore we should consider the second one in the two-body problem with the satellite? What is the corresponding escape velocity?

(b) Assume that the satellite is traveling in such a way that it is moving away from the Earth but always stays 1 AU from the Sun. Plot the

---

[2] You can ignore the fact that the force will vary depending on where on Earth the satellite is, and just calculate this for the average Earth–Sun distance.

magnitude of the gravitational force between the satellite Earth as a function of distance from the Earth and mark (using `plt.axhline`) the gravitational force the Sun is exerting on it. How far from the Earth would the satellite have to travel so the gravitational force on it due to the Earth is the same as the gravitational force due to the Sun? Compare[3] this distance to the Earth–Moon distance (384,000 km) and to the Earth–Sun distance (1 AU)?

(c) Clearly, if the satellite gets further from Earth than in Part 2b, the Sun's influence will be greater than the Earth's (let's ignore the Moon). Suppose now the satellite is 1 AU from the Sun, but further from the Earth than in Part 2b. What velocity does it need to attain now if it is to escape from the gravitational pull of the Sun?

(d) Compare the kinetic energy required to escape from the surface of the Earth (Question 2a) to that required to escape from the solar system (Question 2c).

(e) In the Chapter 5 exercise, we actually calculated the trajectory of an object starting with the escape velocity from a 1 solar mass star starting at 1 AU from that star (the same case as Part 2c!). If we put dimensions appropriate to this problem on that solution (and calculate r from the x- and y-coordinates in that problem), how long does it take (in months) for the satellite to travel so that its distance from the Sun is 4 AU? (Assume no influence from other bodies in the solar system.)

---

[3] I usually intend "compare" to mean "tell me how many times bigger or smaller", i.e., a ratio, not a difference.

An Introduction to Astrophysics with Python
Stars and planets
James Aguirre

Chapter 9

Summary and Commentary

We sum up the state of our treatment of the two-body problem and point ahead to places where we will use the results derived here, in particular in weighing stars and in methods to find exoplanets. We also comment on some features of the two-body problem, which are so special that students should be warned against extrapolating these to more general situations, like interaction with a gravitational potential besides that of a point source, or for cases of many-body interactions.

9.1 The Two-body Problem is Very Special

Given how much time we've devoted to the two-body problem, you may be tempted to think that we've now solved everything of interest in gravitational mechanics. But in many ways, the two-body problem is incredibly special. We were able to obtain a number of exact results, and we can specify everything we might want to know from a few simple equations and knowledge of a few constants, namely the energy and the angular momentum of the orbit (or something from which we can compute those, like the initial conditions). This situation is unusual in mechanics, where the problem is so completely determined by the conservation laws. In fact, for more than two gravitationally interacting masses, the problem is not exactly solvable and indeed, the complexity can be very great.

There are generally two approaches to tackling this problem. One is to still consider the motion of a particle in a gravitational potential, but instead of a $1/r$ potential, we modify the potential to (roughly) describe the effects of all the other particles. The other is to simply let a computer grind away at Newton's equations for all particles and watch what happens.

Quite generically, when you *slightly* perturb the force law away from $1/r^2$, you still get ellipses, but they don't close on themselves, and the orientation of the orbit is said to "precess". This is what happens for planets orbiting the Sun, where most of

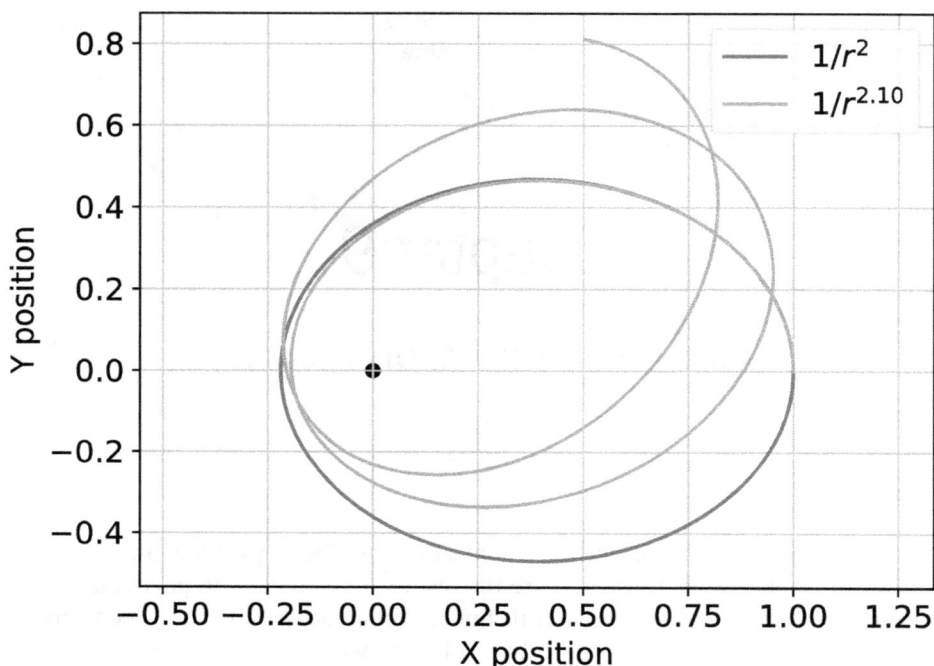

Figure 9.1. An example of an orbit with a force law $1/r^{2+\epsilon}$, with $\epsilon > 0$, leading to a non-closing, "rosette"-like orbit.

the potential comes from the Sun, but there are small perturbations due to the other planets. An example of such orbits (solved using a modified version of the code from Chapter 5) is shown in Figures 9.1 and 9.2. In the case of (trillions of) stars orbiting in a galaxy, the orbits can get considerably more complicated in the "effective potential" of the galaxy.

9.2 Summary of Gravity and Orbits

At this point, it is well worth taking a little time to think about the bigger picture of what we are learning, and how it relates to science more broadly. Here are some thoughts:

1. Newton's Laws provide a description of motion appropriate for many problems, in astrophysics, including the orbits of objects under the force of gravity. In modern form, Newton's Laws specify differential equations for the motion of particles (or objects which can be approximated by them).

2. The notion of physical laws or situations described by differential equations is considerably more general than Newton's Laws, and this language of differential equations for specifying the physics is incredibly useful to know in all branches of science and engineering. (So far, of course, all our differential equations have been derived from Newton's Laws.)

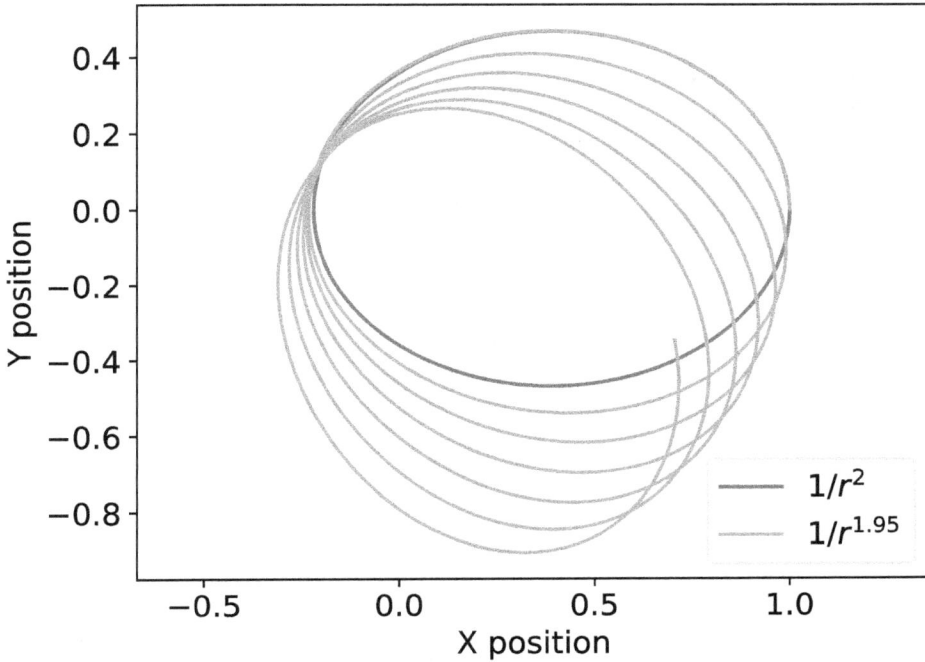

Figure 9.2. An example of an orbit with a force law $1/r^{2+\epsilon}$, with $\epsilon < 0$, again leading to a non-closing, "rosette"-like orbit, though note the change in the sense of the precession.

3. We can solve differential equations in some cases analytically (if we are clever), but in general, we have to appeal to a numerical solution.
4. `Python` and its associated modules provide a powerful tool for numerically solving differential equations.
5. Dimensional analysis and non-dimensionalizing can give us insight into the physics of a problem. For example, The non-dimensional "characteristic" time we introduced in the free-fall problem

$$t_c^2 = \frac{x_c^3}{GM} \tag{9.1}$$

turns out to be related to the period of an orbit: Newton's version of Kepler's Third Law can be stated as "it takes 2π characteristic times to complete an orbit where the characteristic size is the semimajor axis a":

$$P = 2\pi t_c = 2\pi\sqrt{\frac{x_c^3}{GM}} = 2\pi\sqrt{\frac{a^3}{GM}} \tag{9.2}$$

which is exactly Newton's version of Kepler's Third Law:

$$P^2 = (2\pi)^2 \frac{a^3}{GM} \tag{9.3}$$

6. Newton's version of Kepler's Third Law is incredibly powerful and useful, because it allows us to relate observable quantities (period and semimajor axis) to a quantity that is hard to measure otherwise: mass.

7. The two-body problem can be simplified enormously by realizing that motion occurs only in a plane (due to conservation of angular momentum), and the motion can be described in terms of a fictitious problem with the total mass $M = m_1 + m_2$ fixed at the origin, and the reduced mass's motion

$$\mu \equiv \frac{m_1 m_2}{m_1 + m_2} = \frac{m_1 m_2}{M} \tag{9.4}$$

described by the vector **r**, which is actually the separation between the two masses. We can always return to the original problem of tracking each mass relative to the center of mass by finding their position vectors

$$\mathbf{r_1} = -\frac{m_2}{M}\mathbf{r} \tag{9.5}$$

$$\mathbf{r_2} = \frac{m_1}{M}\mathbf{r} \tag{9.6}$$

8. Newton's Laws lead us to the idea of *conserved quantities* like energy and angular momentum, which do not change with time in a problem, and are therefore of considerable help in figuring out general principles to understand the behavior of the solutions even in cases where the exact solution is not available to us—and the two-body problem is such a case, since there isn't an explicit form for $[r(t), \theta(t)]$ or $[x(t), y(t)]$ (though we can solve numerically to arbitrary precision).

9. Energy in the two-body problem can be written many ways depending on what is most useful, but is always kinetic plus gravitational potential. It can be written in a coordinate-system independent way

$$E = \frac{1}{2}\mu \, |\dot{\mathbf{r}}|^2 - \frac{GM\mu}{|\mathbf{r}|} \tag{9.7}$$

in Cartesian coordinates

$$E = \frac{1}{2}\mu\left(v_x^2 + v_y^2\right) - \frac{GM\mu}{\sqrt{x^2 + y^2}} \tag{9.8}$$

in polar coordinates

$$E = \frac{1}{2}\mu(\dot{r}^2 + r^2\dot{\theta}^2) - \frac{GM\mu}{r} = \frac{1}{2}\mu(v_r^2 + v_\theta^2) - \frac{GM\mu}{r}$$

and in terms of an effective potential

$$E = \frac{1}{2}\mu\dot{r}^2 + \frac{L^2}{2\mu r^2} - \frac{GM\mu}{r} = \frac{1}{2}\mu\dot{r}^2 + V_{\text{eff}}(r) \tag{9.9}$$

10. The angular momentum can be written

$$\mathbf{L} = \mu \mathbf{r} \times \mathbf{v} \tag{9.10}$$

with magnitude in both Cartesian and polar coordinates

$$|\mathbf{L}| = \mu(xv_y - yv_x) = \mu r^2 \dot{\theta} \tag{9.11}$$

11. The solution to the two-body problem leads to bound orbits being ellipses (Kepler's First Law) with

$$r(\theta) = \frac{a(1 - e^2)}{1 + e \cos \theta}$$

12. Energy and angular momentum tell us almost[1] everything we want to know about the two-body gravitational problem, and in particular specify the semimajor axis and the eccentricity of the ellipse, for bound orbits, or whether the orbit is bound at all (if $E > 0$).

$$a = \frac{GM\mu}{2|E|} \tag{9.12}$$

$$e = \sqrt{1 + \frac{2EL^2}{G^2 M^2 \mu^3}} \tag{9.13}$$

with the boundary between bound and unbound orbits determining the escape velocity

$$v_{\text{esc}} = \sqrt{\frac{2GM}{r}} \tag{9.14}$$

9.3 Exercise: Orbital Dynamics

Goals for this exercise:
1. To review the many equations describing properties of orbits in the two-body problem
2. To reason conceptually about the two-body problem
3. To interpret equations conceptually rather than numerically

---

[1] For completeness, the thing you can't determine about the orbit just from knowing the energy and angular momentum is the orientation of the ellipse relative to the x–y-axes. We have always assumed that pericenter occurs at $(x_0, 0)$, but this need not be the case. Different initial conditions can give the same angular momentum and energy, but will have the ellipse major axis oriented differently.

In this set of problems, we're going to consider a small satellite of mass m_{sat}, which starts on a *circular* orbit of radius R around the much larger mass m_{earth} of the Earth. As usual, we will define

$$\mu = \frac{m_{\text{sat}} m_{\text{earth}}}{m_{\text{sat}} + m_{\text{earth}}} \tag{9.15}$$

and

$$M = m_{\text{sat}} + m_{\text{earth}} \tag{9.16}$$

We are going to have the satellite perform various "orbital maneuvers", i.e., fire its rockets in various directions, and see what happens. However, firing rockets means applying a non-gravitational forces to the satellite. This violates our basic setup of the two-body problem (only gravity was involved). In order to be able to use the two-body solutions as a guide to the real behavior, in this exercise we'll make some idealized but reasonable assumptions. In particular, we're going to assume

1. The change in mass of the satellite from the ejection of rocket fuel is negligible
2. The time over which the rockets fire is so short that the satellite is in the same *position* in its orbit before and after firing its rockets (i.e., the rocket firing is "instantaneous")
3. Before and after the rockets fire, the physics problem is the two-body problem only between the Earth and the satellite. We will not consider any linear or angular momentum of the rocket exhaust, nor its mass or energy
4. Thus, the effect of firing rockets is to "reset" the initial conditions in the two-body problem to the *same* position, but a new velocity (potentially in both magnitude and direction)

Questions 1 to 5 refer to the following situation.

Suppose that the satellite, initially in a **circular** orbit, fires its rockets in a direction to increase the speed at which the satellite is moving by a factor α (that is, $v_{\text{final}} = \alpha v_{\text{initial}}$, with $\alpha > 1$), which gives rise to a **new** orbit. *The direction the rockets fire is perfectly tangential to the circular orbit.* That is, $\mathbf{v}_{\text{final}} = \alpha v_{\text{initial}} \hat{\boldsymbol{\theta}}$. We will assume that the changes we make by firing the rockets result in the satellite staying in a **bound** orbit ($E < 0$).

In the following, explain your reasoning, using symbolic logic and formulas from the preceding chapters to reinforce your explanations.

Question 1. Does the energy change in going from the circular to the new orbit? Is the magnitude of the energy of the new orbit larger or smaller? You do not need to calculate the energy or its change numerically, but just show symbolically how it is different.

Question 2. Is the semimajor axis of the new orbit larger, smaller, or the same size as previously? Again, give an answer using an equation.

Question 3. Is the angular momentum of the new orbit larger, smaller, or the same size as previously?

Question 4. Is the new orbit still circular? Explain your reasoning.

Question 5. At the moment of the velocity boost is the particle at pericenter (closest approach), apocenter (furthest distance), or some other point in its *new* orbit? Draw a sketch that is consistent with your answers to the previous questions.

Questions 6 to 8 refer to the following situation, illustrated in Figure 9.3. Suppose that the satellite again begins in a circular orbit, but now briefly fires its rockets in a direction perfectly along the line to the Earth, directed so that the radial component of its velocity changes from $v_r = 0$ to $v_r = \beta v_\theta$, where $\beta > 0$. See the figure for the geometry. The velocities are shown immediately *after* the rockets fire, but the orbit is shown as it was immediately *before* the rockets fire.

Question 6. Assuming that the orbit remains bound, is the semimajor axis of the new orbit larger, smaller, or the same size as previously? Explain your reasoning using an equation.

Question 7. Does the angular momentum of the system change? Explain your reasoning using an equation.

Question 8. What is the largest value, $\beta_{\max}$ such that the orbit is still bound? *Hint:* think of the formula for the circular velocity. You should be able to derive a numerical value for $\beta_{\max}$.

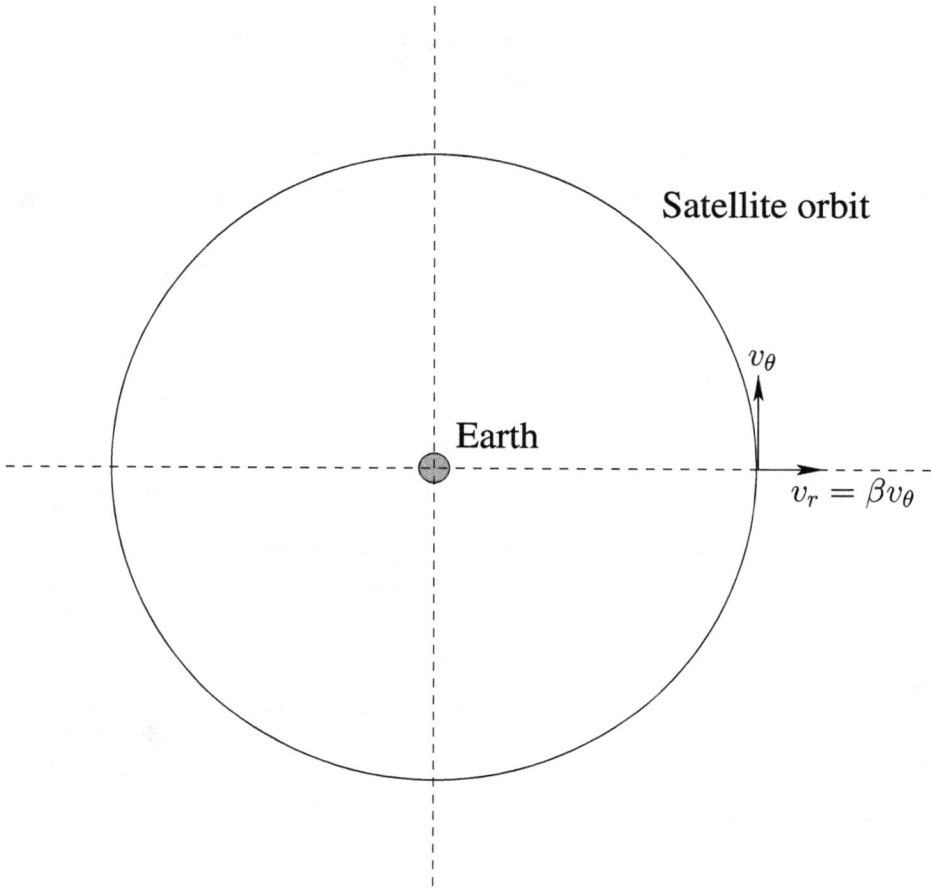

Figure 9.3. The geometry of orbital boosts from a circular orbit. Shown are the initial tangential velocity component v_θ and the additional radial velocity component βv_r.

Let's go back to the circular orbit. We would like to return the satellite to Earth. We consider four possible changes in satellite velocity that might put it on a path to intersect Earth:

1. Increase the tangential velocity ($\alpha > 1$, where α was defined in the problem last time)
2. Decrease the tangential velocity ($\alpha < 1$)
3. Increase the radial velocity away from Earth ($\beta > 0$)
4. Increase the radial velocity toward Earth ($\beta < 0$)

Question 9. Which method will return it to Earth? Explain your reasoning.

Question 10. Now suppose the satellite is in an elliptical orbit. We would now like to boost the satellite out of its bound orbit into an unbound one. We know that the *energy* we need to add doesn't depend on where the satellite is in its orbit. However, the escape velocity, the satellite's kinetic energy, and its gravitational potential energy *do* vary around the orbit. We would like to know if there is a "best" place in the elliptical orbit attempt our escape.

Since mass is at a premium for spacecraft, and we can only carry a finite amount of fuel, we'd like to use it as efficiently as possible. In order to conserve on rocket fuel, we're going to want to apply the smallest possible *force* to the satellite. We can see why this is true using Newton's Third Law, which says the force on the satellite is equal and opposite to the force exerted by the rocket fuel, which is

$$F_{\text{rocket}} = m_{\text{fuel}} a_{\text{fuel}}$$

The acceleration the fuel can achieve is determined by the propellant type and the rocket motor design, so we can regard it as fixed for a given satellite, and thus in order to achieve a larger force we have to burn more fuel m_{fuel}, which pushes toward keeping F_{rocket} as small as possible.

Is there any point in an elliptical orbit at which the force necessary to achieve escape velocity is smallest? Explain your reasoning.

9.4 Study Questions

Here are some additional questions on the first portion of the book.

1. Consider a circular orbit. Write down expressions for the following quantities that may depend on G, M, μ, and R (the radius of the orbit), but *not* on v, the orbital velocity.
   (a) The total energy
   (b) The total angular momentum
   (c) The additional energy required to unbind the orbit

2. A comet of mass 1000 kg is in orbit around our Sun, and has an extremely eccentric orbit of $e = 0.99$ and an orbital period of 70,000 years.
   (a) What is the semimajor axis of the comet's orbit?
   (b) What is the energy of the comet's orbit?
   (c) What is the angular momentum of the comet's orbit?
   (d) What is the escape velocity at closest approach?

3. Consider two planets on elliptical orbits about their Sun. The innermost planet has a semimajor axis a_1 and eccentricity e_1, and the outer planet has semimajor axis a_2 and eccentricity e_2. Their perihelion points are both aligned to occur at $\theta = 0$. (See Figure 9.4 for an example with $a_1 = 1$, $e_1 = 0.8$, $a_2 = 3$, $e_2 = 0.2$.) Assume $a_1 < a_2$ and $e_1 > e_2$. In terms of a_1, e_1, a_2, and e_2, how far apart are the planets when both are at perihelion? Aphelion?

4. Captain Picard brings the USS Enterprise to a new planet and enters an elliptical orbit of it with eccentricity $e = 0.3$. He notes that it takes them 2

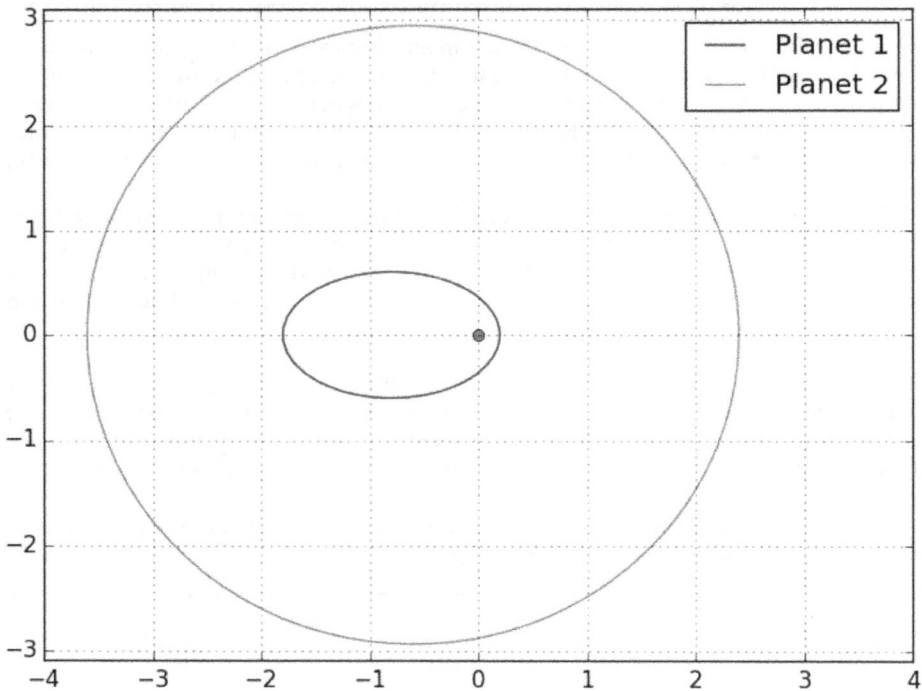

Figure 9.4. Two planetary orbits with their semimajor axes aligned.

hours to go around the planet, and their closet point is 7000 km from the planet's *center*.

(a) Without knowing the mass of the Starship Enterprise, estimate the mass of the planet.

(b) Captain Picard asks Engineering what the current mass of the ship is, and they respond 4.6×10^8 kg. How much error (fractionally) did you make in the mass of the planet by ignoring the mass of the Enterprise?[2]

5. For this problem, use Kepler's Third Law in the form applicable to our solar system

$$P^2 = Ka^3$$

with $K = 1 \, \text{year}^2/\text{AU}^3$ and assume all planets travel in perfectly circular orbits.

(a) Find the orbital velocity v as a function of the planets orbital semimajor axis a. In words, what is the trend of the planet's velocity as its distance from the Sun increases?

(b) Calculate the velocity of Saturn in km s^{-1} (again assuming a circular orbit), *using the result from the first part*. The semimajor axis of Saturn's orbit is $a = 9.6$ AU.

---

[2] How could we know that, you ask? http://goo.gl/A886JM

(c) Astronomers like km s^{-1}, as typical values of stellar velocities (for example) are in the few to 100's of km s^{-1}. How many miles per hour is this? (For those who would like, `from astropy.units import imperial` will give you access to `imperial.mile`.)

6. **The formation of stars from gravitational collapse.**[3] A simplified model of the gas cloud from which the Sun collapsed (the "solar nebula") was that is was uniform density sphere of mass equal to the Sun's mass (2.0×10^{30}kg), with a radius of approximately one light-year (1 ly $\approx 9.5 \times 10^{15}$m).

   (a) The *number density n* of particles in this cloud is defined by

   $$n = \frac{1}{m_{particle}} \frac{M}{V} = \frac{1}{m_{particle}} \rho \qquad (9.17)$$

   where ρ is usual mass density (mass M per volume V) and $m_{particle}$ is the mass of one particle, which we will assume to hydrogen, with a mass of approximately $m_H \approx 1.7 \times 10^{-27}$kg.

   i. What are units of number density?

   ii. What is the number density of the solar nebula, given the numbers above?

   iii. Compare this number density to the best vacuum I've ever made in my lab, with pressure $P \approx 1 \times 10^{-3}$ N m^{-2} (1 N m^{-2} = 1 Pascal (Pa)) at temperature $T = 300$ Kelvin

   $$n = \frac{P}{kT} \qquad (9.18)$$

   where $k = 1.38 \times 10^{-23}$ J K^{-1} is the Boltzmann constant.

   (Hey! Sneaky—that's the ideal gas law up there!)

   Note: "Compare" almost always means ratios: how many times bigger or smaller. Absolute differences are rarely helpful when comparing quantities that can differ by huge amounts.

   (b) The time for the gas cloud to collapse under the force of gravity can be approximated using the free fall under a $1/r^2$ potential from Exercise I.3 for a particle with an initial position in the outermost part of the cloud, and feeling the full gravitational attraction of the whole cloud (i.e., its total mass). How long is this free-fall time for the solar nebula? Express your answer in years.

7. In this chapter's exercise, we changed the velocity of a circular orbit by increasing the tangential velocity to αv_{circ} and we added a velocity in the radial direction of magnitude βv_{circ}. In both cases, the resulting orbits are no longer circular. What are the eccentricities, written in terms of α, G, M, and R (the circular orbit radius) and β, G, M, and R for the two cases?

8. Derive expressions for the pericenter and apocenter velocities of an elliptical orbit of eccentricity e in terms only of e and the v_{esc} at that position.

---

[3] And a chance to practice using `astropy` units.

This series of questions is about landing on, and leaving, the surface of the Moon. Some questions do depend on correctly answering other questions; I have tried to minimize this, and where I could not, I have indicated it. If you have trouble answering a question on which another depends, indicate how, if you had solved the previous problem, your new answer would depend on that answer, and assume a value that allows you to proceed.

You may use the following information and make the following assumptions.

- The mass of the Moon is $m_{\mathrm{moon}} = 7.34 \times 10^{22}$ kg.
- It is a perfect sphere with radius 1740 km.
- Since the Moon has no atmosphere, we need not consider complications like air resistance.
- You can ignore the gravitational effects of the Moon and Sun and just consider the gravitational interaction between your spacecraft and the Moon.
- Your spacecraft has a mass of $m_{\mathrm{spacecraft}} = 1 \times 10^4$ kg.
- Several orbits are shown in a plane around the Moon in Figure 9.5. The surface of the Moon is shown in gray, and an initial circular orbit with height above the Moon's surface $h_{\mathrm{orbit}} = 100$ km is shown in blue.
- For all the questions where you change your orbital motion, we assume the starting position is as given in Question 9.5 by the point indicated with a blue star, at $x = -R_{\mathrm{circ}}$, $y = 0$. The sense of the orbits is *clockwise*, that is, starting from the blue star, proceed up and to the right around the orbit.

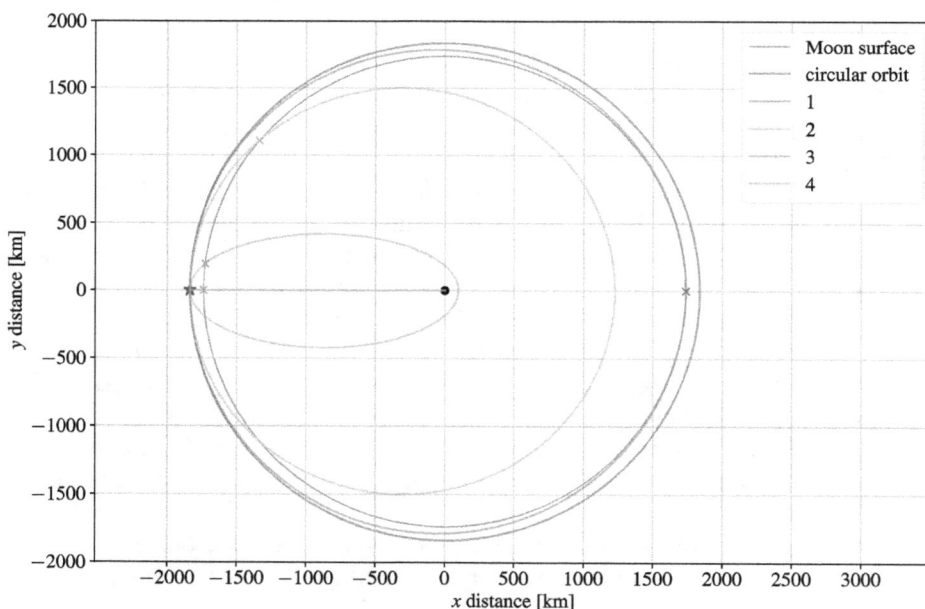

Figure 9.5. A plane view of the surface of the Moon, an initial circular orbit, and four other orbits around the Moon's center of mass which intersect the Moon's surface.

- Four other orbits are shown, labeled 1 to 4. All of them have Moon's center of mass (black dot) at one focus, and the initial position of your spacecraft marked with a blue star. The point of intersection of the orbit with the Moon's surface is marked with an X.
- Similarly to this chapter's exercise when our spacecraft changes its orbit from the circular one, we're going to assume that:
  - the change in mass of the spacecraft from the ejection of rocket fuel is negligible
  - the time over which the rockets fire is so short that the spacecraft is in the same *position* in its orbit before and after firing its rockets (i.e., the rocket firing is "instantaneous")
  - before *and* after the rockets firing, the problem is the two-body problem only between the Moon and the spacecraft. We will not consider any energy, linear or angular momentum of the rocket exhaust

1. What are the orbital speed, energy, and angular momentum of the circular orbit?
2. Suppose you had enough fuel to completely stop your motion in the circular orbit around the Moon, so that now $|\mathbf{v}| = 0$. How much time would it take to fall to the Moon's surface? Use the approximation of a constant force, generating a constant acceleration, given by the gravitational acceleration at your initial position. Remember, you are only falling a distance h_{orbit}. For definiteness, this is the orbit marked 4 in Figure 9.5.
3. ⋆ What is the percentage difference between the answer in 2 and the time to fall using the proper $1/r^2$ force law in place of the constant force? (This is a callback to Chapter 4. One advantage of the dimensionless solution approach is that you need only change the values of mass and initial position in exercise in that chapter.)
4. How much does your orbital *energy* change if you completely bring your velocity to zero as in Part 2? Be careful with the sign of the energy. What is your angular momentum after this change?
5. For all of the orbits that intersect the surface of the Moon, what is the common feature of your starting point (the blue star)? What does this imply about the orientation of your velocity in the new orbits 1–4 at the starting point?
6. Now suppose you want to change your orbital velocity to enter an elliptical orbit which will *just* intersect the surface of the Moon, labeled in Figure 9.5 as orbit 1. What is the semimajor axis of this new orbit? What is the eccentricity of the orbit? *Hint*: this problem just uses geometry.
7. How much do you need to change your kinetic energy to go from the circular orbit to orbit 1? What is the orientation and magnitude of your new velocity? The answer here does depend on getting the answers to Questions 5 and 6 correct, so think this through.
8. Using the relation for the energy of a bound orbit in terms of the semimajor axis and the energy in terms of the kinetic and potential energies, derive a

relation for the magnitude of the velocity of that orbit as a function of the separation distance r and the semimajor axis a (as well as G and M). If you find yourself doing more than a couple of lines of algebra, something has gone wrong.

9. Use the numerical integration methods from class to start at the initial point in orbit 3, which has an eccentricity of 0.9 and a semimajor axis of 967 km. You will need to find the initial velocity for this orbit (there are a couple of methods given the problem statement). What is your radial velocity v_r toward the center of the Moon and your tangential velocity v_θ across the surface of the Moon just before you land? (You can check that the magnitude of your velocity agrees with the prediction of the equation from Question 8.)

10. Having landed on the Moon, you now wish to return to Earth. Sitting on the surface of the Moon, what escape velocity do you need?

Part II

Interaction of Light and Matter

An Introduction to Astrophysics with Python
Stars and planets
James Aguirre

Chapter 10

Light and the Electromagnetic Spectrum

We describe the centrality of the study of light for astronomical observations. We discuss basic properties of light, both in terms of its wave-like and particle-like properties. We discuss the electromagnetic spectrum, some ideas of the interaction of light and matter, and the basic idea of a spectrum produced under various conditions of astrophysical interest.

10.1 The Importance of Light

As astronomers, we are keenly interested in the properties of light, as it is the primary means by which we get information about distant objects. While we can now detect gravitational radiation from various collisions in the universe, and we can detect neutrinos, as well as cosmic rays (high-energy particles) from some objects, the bulk of the information we can learn still comes from electromagnetic radiation, which I will often refer to generically as "light".[1] A beautiful example of the rich amount of information provided by light is shown in Figure 10.1.

10.2 Properties of Light

The classical description of electromagnetic radiation is via Maxwell's equations, which describe the behavior of electric $\mathbf{E}$ and magnetic $\mathbf{B}$ fields, and have solutions that include a wave propagating in free space. For an electromagnetic wave propagating in free space, the electric field $\mathbf{E}$ is perpendicular to the magnetic field $\mathbf{B}$, which in turn is perpendicular to the direction of propagation. This is illustrated in Figure 10.2.

---

[1] For our purposes, we will treat "light", "electromagnetic waves", "photons" or "radiation" as synonymous, even when we are not talking about visible light.

doi:10.1088/2514-3433/ade5f6ch10

Figure 10.1. A tremendous amount of information is encoded in the light we receive from astronomical objects. This information comes from the spectrum and angular distribution of the emission, and from both the emitting material and about what it passes through on its way to us. Shown is an image of the Crab Nebula, the remnants of a star that exploded in 1054, taken by the James Webb Space Telescope (JWST). The different colors in the image correspond to different wavelengths in the infrared. Image credit: NASA, ESA, CSA, STScI, Tea Temim (Princeton University)

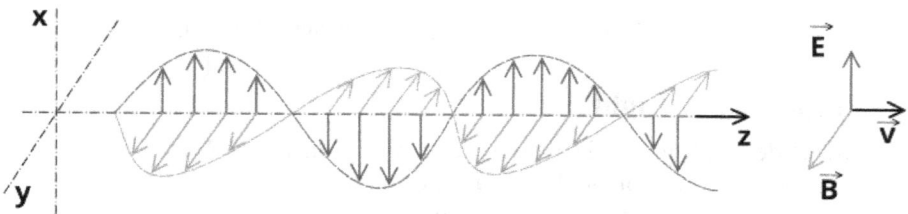

Figure 10.2. A visualization of electromagnetic waves. Note that electric and magnetic fields shown are only those along the indicated line at a given moment in time. The extent of the fields along the transverse dimensions is just to show the relative strength of the field and the orientation of the vector. "This [Onde electromagnetique] image has been obtained by the author(s) from the Wikimedia website where it was made available by [Chetvorno] under a CC BY-SA 3.0 licence. It is included within this article on that basis. It is attributed to [SuperManu]." https://creativecommons.org/licenses/by-sa/3.0/

These waves (in a vacuum) travel at a constant *speed of light c*, independent of their frequency:

$$c = 299,\ 792,\ 458 \text{ m s}^{-1} \tag{10.1}$$

Note that this definition of the speed of light is *exact*: by defining the speed of light this way and defining the second as the duration of a specific number of cycles of a light wave[2], we *define* the meter. (The meter was historically defined by various other standards.)

For light waves, the frequency is related to the wavelength (in vacuum) by

$$\boxed{c = \lambda\nu} \tag{10.2}$$

The wavelength is given in meters. The wavelength of visible or ultraviolet light is also quoted in angstroms ($1 \text{ Å} = 1 \times 10^{-10}$ m) and nanometers ($1 \text{ nm} = 1 \times 10^{-9}$ m $= 10$ Å).

The rate of energy transfer in the direction that the light wave is traveling is given by

$$\mathbf{S} = \frac{1}{\mu_0}\mathbf{E} \times \mathbf{B} \tag{10.3}$$

and the units of $\mathbf{S}$ are W m^{-2}, i.e., flux. The constant μ_0 is the *magnetic constant*

$$\mu_0 = 4\pi \times 10^{-7}\frac{\text{Vs}}{\text{Am}} \tag{10.4}$$

where V is Volts and A is Amperes. It is one of three constants that appear in Maxwell's equation for electromagnetism. The second is c, the speed of light, and the final one is ε_0, which is related to these two by

$$\varepsilon_0\mu_0 = \frac{1}{c^2} \tag{10.5}$$

In free space, we have $E_0 = cB_0$, so, averaging over one cycle of the wave, we have

$$\langle S \rangle = \frac{E_0^2}{2c\mu_0} \tag{10.6}$$

This is about as much as we will say about classical light waves in this book because it turns out that light also has particle-like properties, and these will turn out to be more important in the kinds of interactions with matter we are going to be interested in. These particle-like properties were first discovered about the time that people were trying to puzzle out the behavior of atoms. We refer to these particles of light as *photons*. Photons are strange things: they can act individually like *particles*, but collectively they act like *waves* (and yet they do not interact directly with each other). You can imagine a photon as something that looks like Figure 10.3, termed a *wave* packet.

Note that the wave packet is (somewhat) localized in space and time (like a particle), but oscillates like a wave. (Classically, the amplitude of the oscillation is related to the electric field.)

---

[2] Specifically, "the duration of 9,192,631,770 periods of the radiation corresponding to the transition between the two hyperfine levels of the ground state of the cesium 133 atom."

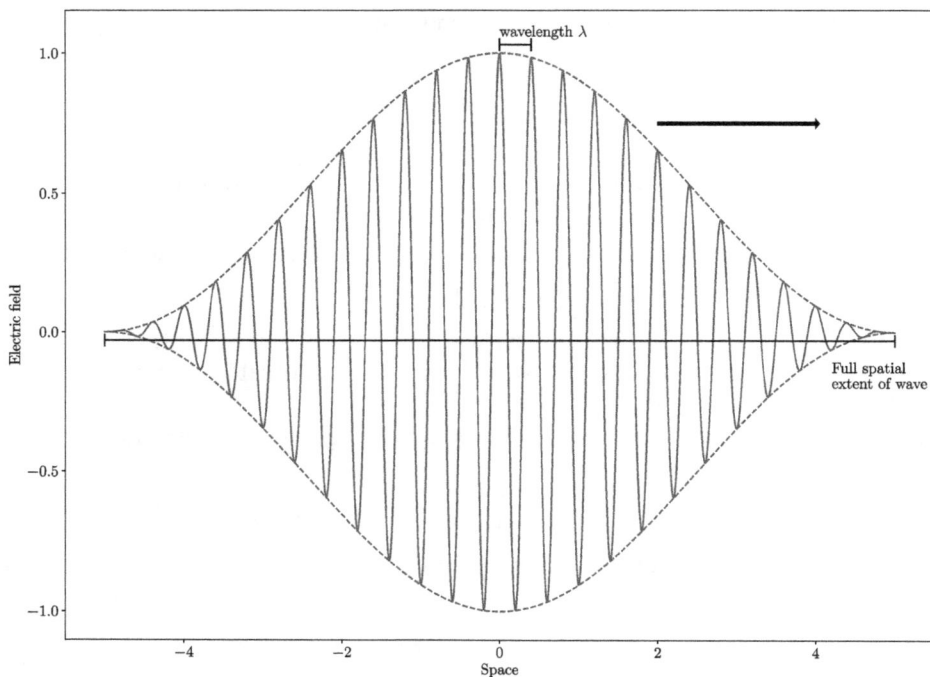

Figure 10.3. A wave packet, showing both the characteristic oscillation frequency, and the localization of the packet in space. With a re-labeling of the x-axis, we can also think of wavepackets localized in time as well.

Photons have two basic properties:
1. A photon is uniquely characterized by its frequency of oscillation ν. The frequency is quoted in Hz (s^{-1}) and its SI multiples (MHz, GHz, THz, etc.).
2. All photons travel at the speed of light c

Corresponding to their particle-like properties, each photon carries an energy

$$\boxed{E = h\nu} \tag{10.7}$$

where h is *Planck's constant*

$$h = 6.626 \times 10^{-34}\,\text{J s} = 2\pi \tag{10.8}$$

Note that the units of Planck's constant are the same as angular momentum. Note that λ, ν, and E are all related to each other for light.

Photons do not have mass.[3] Note that, unlike a massive particle, the photon's energy has nothing to do with its speed (which is always c). Similarly, we will not be

---

[3] Again, many of you will know that gravity can bend the direction of travel of light, because Einstein's gravity interacts with mass *and* energy (which are the same). The more precise statement is that photons do not have rest mass, i.e. an energy not associated with their frequency of oscillation.

able to relate the photon's momentum to its speed, but nevertheless, photons *do* have momentum:

$$p = \frac{h}{\lambda} = \frac{h\nu}{c} = \frac{E}{c} \qquad (10.9)$$

Because photons carry momentum, if that momentum changes in interacting with matter, then they can exert a *force* (Newton's second law), and thus also a pressure (force per unit area). This pressure is given by

$$P = \frac{\Phi}{c} \qquad (10.10)$$

where Φ is the flux of photons (in units of W m^{-2}). We will have more to say about photon fluxes later in the course.

Photons are one of the three commonly occurring ways of transporting energy from one place to another, the other two being conduction (thermal contact) and convection (bulk motion of matter). We will come back to this topic when we talk about the interiors of stars.

10.3 The Electromagnetic Spectrum

The only thing that distinguishes various forms of electromagnetic radiation from one another is their wavelength (or frequency); otherwise, their mathematical description is identical.

However, light of different wavelengths will interact with matter in very different ways, and so we distinguish the different wavelengths with different names:

Table 10.1. The Electromagnetic Spectrum, with Various Regions in Wavelength, Their Common Names, and Typical Processes Generating the Radiation.

Wavelength Range	Name	Typical Physical Processes
$\leqslant 1$ nm	γ (gamma)-ray	nuclear reactions, very high-energy collisions
1–10 nm	X-ray	inner shell electrons of atoms, high-energy collisions
10–400 nm	ultraviolet	electron transitions in atoms
400–700 nm	visible	electron transitions in atoms
700 nm–1 mm	infrared	thermal radiation, vibrations in molecules
1 mm–10 cm	microwave	rotational motion of molecules
$\geqslant 10$ cm	radio	thermal and non-thermal motions of free electrons

Example Blue material. An object illuminated by light of many colors (white), which *reflects* blue light and *absorbs* other colors will appear to us as blue.

Example Water molecules.
- do not interact much with optical light (which is why water is transparent!)
- absorb UV and shorter wavelengths, as well as IR
- absorb photons with $\lambda = 12.24$ cm very strongly (microwave oven!)

There are also materials that are capable of generating photons from internal energy (heat, nuclear, etc.), e.g., a light bulb, burning coal, stars, etc. The *spectrum* of the object is the number of photons (or power) emitted as a function of wavelength (or frequency), and in general the observed spectrum is a complicated function depending on many variables, such as the chemical composition of the material and the details of its interaction with the surrounding medium. Some luminous objects will emit photons across a wide range of wavelengths, that is, a *continuous spectrum*. For sufficiently dense materials, this emission will approach an ideal where the properties of the spectrum depend only on the temperature; we will discuss this further in Chapter 14 on thermal emission. By contrast, a diffuse gas of atoms or molecules is observed to emit an *emission line spectrum* of only certain well-defined wavelengths (corresponding to photon energies). Of particular interest to us as astronomers is the *absorption line* spectrum, where a diffuse gas is in front of a source of continuum emission. This can be a physically distinct gas cloud through which the light passes on its way to us, or it can be closely associated with the source of continuum radiation. An important example is the atmosphere of a star or planet, with the warm interior of the star or the warm surface of the planet forming the source of continuum emission passing through the diffuse gas of the atmosphere. The important topic of absorption lines in the atmospheres of stars are the subject of Chapter 22.

An Introduction to Astrophysics with Python
Stars and planets
James Aguirre

Chapter 11

An Introduction to Atoms

Here we take many of the ideas developed in the two-body problem and apply them to the interaction of a proton and electron, resulting in the model of a hydrogen atom, the simplest and most important of the atoms in astrophysics. We look at the condition for charged particles to radiate and introduce the Larmor formula. The fact that our classical picture of orbits and radiation leads to unstable atoms that emit continuously, instead of at the observed special wavelengths, provides an indication that, while concepts like energy and angular momentum will survive in our description of atoms, some modifications are required.

Because we are interested in how light interacts with matter, we are going to need some machinery in place to understand this quantitatively. We're going to begin with a simple model of hydrogen—an important astrophysical element, as it is the most abundant—which is just a proton and electron bound together.

11.1 A Model of Matter

Having put so much effort into the two-body problem, let's see if we can apply it other places. One idea of a place to look is to recall that, mathematically, the force between two *charged* particles interacting via the electrostatic force looks very similar to the gravitational force. If q_1 and q_2 are the charges of two particles, then the force between them is

$$\mathbf{F} = \frac{1}{4\pi\epsilon_0}\frac{q_1 q_2}{r^2}\hat{r} \tag{11.1}$$

where r is the distance between the two particles and $\hat{r}$ is the direction of the line connecting them. The charges q_i are measured in SI units in Coulombs (C), and it turns out to be a fact of nature that electric charge only comes in integer multiples of a fundamental constant $e = 1.6022 \times 10 - 19$ C. The units of ϵ_0 must be

doi:10.1088/2514-3433/ade5f6ch11

$$[\epsilon_0] = \left[\frac{C^2}{Nm^2} \right] \tag{11.2}$$

and numerically it is $\epsilon_0 = 8.8542 \times 10 - 12$. Unlike mass, which only comes in a positive variety (and for which the force law is always attractive), charges come in both positive and negative flavors, and the force will be either repulsive for like charges or attractive for dissimilar ones.

Thus, if we describe a new two-body problem as two particles of opposite charges moving under the influence of their electrostatic attraction, we can carry over most of the math of the gravitational two-body problem, making the substitution

$$Gm_1m_2 \rightarrow \frac{q_1q_2}{4\pi\epsilon_0} \tag{11.3}$$

Since the hydrogen atom is made up of one proton (charge $+e$) and one electron (charge $-e$), this is the problem we will consider. We note that $m_p \approx 2000m_e$, so the masses are also very unequal, much like the solar system problem. Happily for us, hydrogen is the most abundant element in the universe, so it's worth studying it in a little detail.

In the exercises, we will answer the question of whether we need to include the gravitational interaction between the proton and electron in our formulation, by comparing the strength of the gravitational attraction to the electrostatic force between a proton and electron. For now, we will proceed by ignoring the gravitational force in the hydrogen atom problem.

11.2 The Classical Hydrogen Atom

To see what changes by changing the constants in the force law, let's go back to the original problem, which now looks like:

$$m_1\ddot{\mathbf{r}}_1 = -\frac{e^2}{4\pi\epsilon_0} \frac{1}{|\mathbf{r}_1 - \mathbf{r}_2|^2} \frac{\mathbf{r}_2 - \mathbf{r}_1}{|\mathbf{r}_1 - \mathbf{r}_2|}$$

$$m_2\ddot{\mathbf{r}}_2 = -\frac{e^2}{4\pi\epsilon_0} \frac{1}{|\mathbf{r}_1 - \mathbf{r}_2|^2} \frac{\mathbf{r}_1 - \mathbf{r}_2}{|\mathbf{r}_1 - \mathbf{r}_2|}$$

Again, we can choose to go to the center of mass frame, defined by

$$m_1\mathbf{r}_1 + m_2\mathbf{r}_2 = 0 \tag{11.4}$$

and use the definition of the vector connecting the two masses

$$\mathbf{r} = \mathbf{r}_2 - \mathbf{r}_1 \tag{11.5}$$

to eliminate $\mathbf{r}_1$ and $\mathbf{r}_2$ in favor of $\mathbf{r}$. Dividing by the total mass $M = m_1 + m_2$

$$\frac{m_1}{M}\mathbf{r}_1 + \frac{m_2}{M}\mathbf{r}_2 = 0 \tag{11.6}$$

so

$$\frac{m_1}{M}\mathbf{r_1} + \frac{m_2}{M}(\mathbf{r_1} + \mathbf{r}) = 0 \tag{11.7}$$

$$\mathbf{r_1} = \frac{-m_2}{M}\mathbf{r} \tag{11.8}$$

$$\mathbf{r_2} = \frac{-m_1}{M}\mathbf{r} \tag{11.9}$$

Now, the electrostatic force has an associated potential energy of the same form as the gravitational one, so we can write the total energy of our hydrogen atom as

$$E = \frac{1}{2}m_1\dot{\mathbf{r}}_1^2 + \frac{1}{2}m_2\dot{\mathbf{r}}_2^2 - \frac{e^2}{4\pi\epsilon_0 r} \tag{11.10}$$

$$E = \frac{m_1 m_2^2}{2M^2}\dot{\mathbf{r}}^2 + \frac{m_2 m_1^2}{2M^2}\dot{\mathbf{r}}^2 - \frac{e^2}{4\pi\epsilon_0 r} = \frac{1}{2}\mu\dot{\mathbf{r}}^2 - \frac{e^2}{4\pi\epsilon_0 r} \tag{11.11}$$

and, again as before, we can re-write this for a single scalar function $r(t)$

$$E = \frac{1}{2}\mu\dot{r}^2 + \frac{L^2}{2\mu r^2} - \frac{e^2}{4\pi\epsilon_0 r} \tag{11.12}$$

The solutions will be ellipses (or circles) for $E < 0$, and we can write down expressions for the semimajor axis and eccentricity of the orbit depending, as in the gravitational case, on both $|E|$ and L.

As good as all this is, it was realized that this description of the hydrogen atom had certain problems, which the gravitational problem did not. The essence of the problem was that while in the so-called *classical mechanics* of Newton, the quantities E and L could be anything, set only by the initial conditions, this kind of assumption didn't work when talking about atoms. For one thing, while the size of the solar system could have been different depending on its initial conditions, hydrogen atoms are always the same size: somehow E is not completely arbitrary.

11.3 Matter and Radiation

Now we need to tie together light and electric charges. The important principle that you will learn in an electricity and magnetism course is that accelerated electric charges do something that accelerated masses do not: they *radiate*.[1] That is, there is more to electromagnetism than the electrostatic (Coulomb) force, and we have not included it in our description of the problem. In that E&M course, you will probably

---

[1] Some of you may be aware that indeed, in Einstein's theory of gravity, accelerated masses *do* radiate so-called gravity waves. However, for speeds low compared to the speed of light and for small-ish masses, these effects are negligible.

derive the so-called Larmor formula, which says that an accelerated charge radiates a power (Joules per second, or Watts) of

$$P = \frac{2}{3}\frac{e^2}{4\pi\epsilon_0}\frac{a^2}{c^3} \tag{11.13}$$

where a is the acceleration (not a semimajor axis). For the circular orbit of an electron, its acceleration is given by

$$a = \frac{v^2}{r} = \frac{F}{m_e} = \frac{e^2}{4\pi\epsilon_0 m_e}\frac{1}{r^2} \tag{11.14}$$

constant in magnitude, and always perpendicular to the velocity.

11.4 Exercise: Atoms and Radiation

11.4.1 Goals

- Understand the units for electrostatic forces
- Understand the relation of the classical hydrogen atom to the gravitational two-body problem
- Understand about radiation from charges and the problem with the classical picture of atoms

The relevant electrostatic constants are available in astropy as

```
print(c.eps0) # epsilon_0
print()
print(c.m_p) # proton mass
print()
print(c.m_e) # electron mass
print()
print(c.e.si) # elementary charge magnitude in SI units (of
    proton or electron)
```

```
Name = Vacuum electric permittivity
Value = 8.8541878128e-12
Uncertainty = 1.3e-21
Unit = F/m
Reference = CODATA 2018
```

```
Name = Proton mass
Value = 1.67262192369e-27
Uncertainty = 5.1e-37
Unit = kg
Reference = CODATA 2018
```

```
Name = Electron mass
Value = 9.1093837015e-31
Uncertainty = 2.8e-40
Unit = kg
Reference = CODATA 2018
```

```
Name = Electron charge
Value = 1.602176634e-19
Uncertainty = 0.0
Unit = C
Reference = CODATA 2018
```

We can also get additional information about units

u.A.long_names

['ampere', 'amp']

u.A.physical_type

PhysicalType('electrical current')

Question 1. What is the long_names and physical_type for u.C? What units does astropy say the units of ϵ_0 are? Explain how these can be converted to the units given in Equation (11.2). Use Python to demonstrate the conversion and show the answers in your notebook.

Question 2. What is the ratio of the electrostatic force between a proton and an electron to the gravitational force between them, if they are separated by a distance $r = 1$ m? First work out algebraically what the ratio should be below and then do the calculation numerically.

Question 3. The electrostatic force has an associated potential energy of the same form as the gravitational one, so we can write the total energy of the hydrogen atom as

$$E = \frac{1}{2}\mu|\mathbf{v}|^2 - \frac{e^2}{4\pi\epsilon_0 r} \tag{11.15}$$

In analogy with the gravitational two-body problem, find expressions for the semimajor axis of an electron orbit around a proton under the influence of the electrostatic force in terms of $|E|$, and for the radius of a *circular* orbit in terms of L.

Question 4. The equations above won't tell you what the energy (or size) for an atom *should* be, because E and L are set by the initial conditions, and there are a wide range of possibilities for both consistent with a bound orbit. But atoms do in fact have a characteristic size, which is the same for all atoms of a given element. For hydrogen, this is about $r_0 = 5 \times 10^{-11}$ m. If we take the circular orbit radius of an electron around a proton to be r_0, what is the corresponding energy of the orbit (in Joules)? Use your result from Question 3 and calculate using Python.

Question 5. How much power does an electron orbiting at a radius r_0 emit (in Watts)? Use the Larmor formula. Recall that

$$a = \frac{v^2}{r} \tag{11.16}$$

for a circular orbit. Write Python functions to calculate the acceleration given the radius, and the power given the acceleration.

Question 6. Assuming the electron doesn't change the radius of its orbit, how long can it go on radiating this power? Explain your reasoning for how to answer this, and then use Python to calculate the answer.

Question 7. If an electron in a circular orbit makes a single orbit at fixed radius, radiating away some of its kinetic energy, what must happen to its semi-major on the next orbit? Explain your reasoning.

An Introduction to Astrophysics with Python

11-6

An Introduction to Astrophysics with Python
Stars and planets
James Aguirre

Chapter 12

Atoms and Atomic Processes

To solve the problems of the classical model of the atom of both the atom's instability and the fact that it does not emit a line at a well-defined wavelength, we introduce the necessary quantum-mechanical ideas of angular momentum and energy quantization and provide a simple description of the hydrogen atoms. We go on to consider how to build up the structure of more complicated atoms. We also consider what quantum mechanics has to say about the exchange of energy between atoms in collisions and in interaction with light. This description includes collisional excitation and de-excitation and radiative processes (spontaneous and stimulated emission, absorption).

We now introduce the kinds of things atoms, photons, and electrons can do.

12.1 Transitioning from Classical to Quantum Hydrogen

The classical theory of electromagnetic radiation says that the accelerated electron going around the proton will radiate. If this accelerated electron radiates and the energy of the electron could be anything, then the energies of the associated frequencies (and wavelengths) of the emitted light could be anything at all. However, when scientists (and we) looked at the light coming from a pure gas of hydrogen, they did *not* see all possible colors: rather, only certain very narrow ranges of wavelength showed much emission, and most other wavelengths were not emitted. Based on what we know about light, this suggested that only certain very specific *energies* of light were emitted, and so one proposal was that instead of every possible orbit being allowed, for some reason only certain orbital energies were permitted.

doi:10.1088/2514-3433/ade5f6ch12

There are a number of ways of varying degrees of sophistication to think about the problem, but in the early days of the 20th century people reasoned as follows: suppose that, just like light, electrons should also have a wavelength associated with them, that was related to their momentum

$$p = \frac{h}{\lambda} \tag{12.1}$$

Then, when an electron is traveling in free space, its wavelength can be any real number, but if it is forced to travel in a circular orbit, it must fit an integer number of wavelengths around the circle, that is,

$$2\pi r_c = n\lambda \tag{12.2}$$

where n is an integer $n = 1, 2, 3, \dots$. Then the angular momentum associated with this circular orbit is

$$L = r_c p = r_c \frac{h}{\lambda} = r_c \frac{h}{2\pi r_c / n} = \hbar n \tag{12.3}$$

where we've introduced $\hbar = h/(2\pi)$. Let's now go back to our equations for the circular orbits of electrons and protons going around each other

$$a = \frac{e^2}{8\pi\epsilon_0 |E|} \tag{12.4}$$

and, for *circular* orbits

$$r_c = \frac{4\pi\epsilon_0}{e^2\mu} L^2 \tag{12.5}$$

and re-write them in terms of this *quantized L* in Equation (12.3)

$$r_c = \frac{4\pi\epsilon_0}{e^2\mu} \hbar^2 n^2 \equiv a_0 n^2 \tag{12.6}$$

where

$$a_0 = \frac{4\pi\epsilon_0 \hbar^2}{e^2\mu} = 5.29 \times 10^{-11} \, \text{m} \tag{12.7}$$

is a particular length scale, usually referred to as the *Bohr radius*. Then, for a circular orbit

$$|E| = \frac{e^2}{8\pi\epsilon_0} \frac{1}{a_0 n^2} \tag{12.8}$$

or

$$|E| = \frac{e^2}{8\pi\epsilon_0} \frac{e^2\mu}{4\pi\epsilon_0 \hbar^2} \frac{1}{n^2} = \left(\frac{e^2}{4\pi\epsilon_0 \hbar} \right)^2 \frac{\mu}{2n^2} \tag{12.9}$$

12-2

It turns out that the combination

$$\alpha \equiv \frac{e^2}{4\pi\epsilon_0 \hbar c} \approx \frac{1}{137} \tag{12.10}$$

is dimensionless, and thus the energy can be rewritten as

$$|E| = \frac{m_e c^2}{2}\alpha^2 \frac{1}{n^2} \tag{12.11}$$

where I've made the approximation $m_e \approx \mu$ for this problem.

This leads to a really remarkable result: *discrete energy levels* in the hydrogen atom, with energies given by

$$\boxed{E_n = -\frac{m_e c^2}{2}\alpha^2 \frac{1}{n^2}} \tag{12.12}$$

where $n = 1, 2, 3, \ldots$. That is, the electron can't just have any energy, but must have one determined by the integer n. Note that the fact that $n = 1$ is the minimum value means that there is a definite size associated with a hydrogen atom, namely $a_0 = 5.29 \times 10^{-11}$ m.

The quantity

$$\mathrm{Ry} = \frac{m_e c^2}{2}\alpha^2 = 2.18 \times 10^{-18}\mathrm{J} \tag{12.13}$$

has sort of impractically small units, so we often define

$$1\mathrm{eV} = 1.6022 \times 10^{-19}\mathrm{J} \tag{12.14}$$

so that

$$\mathrm{Ry} = 13.6 \quad \mathrm{eV} \tag{12.15}$$

12.2 Radiation from the Hydrogen Atom

If we argue that electrons can only be in one of an (infinite number) of discrete energy levels, then we can ask what happens when the electron changes from a higher energy level (closer to zero energy) to a lower one. This answer turns out to be related to the notion of an accelerated particle radiating.

In an electricity and magnetism course, you will derive the so-called Larmor formula, which says that an accelerated charge radiates a power (Joules per second, or Watts) of

$$\boxed{P = \frac{2}{3}\frac{e^2}{4\pi\epsilon_0}\frac{a^2}{c^3}} \tag{12.16}$$

For the circular orbit of the electron, its acceleration is given by

$$a = \frac{v^2}{r} = \frac{F}{m_e} = \frac{e^2}{4\pi\epsilon_0 m_e}\frac{1}{r^2} \tag{12.17}$$

constant in magnitude, and always perpendicular to the velocity. Using the circular radius corresponding to n, we get

$$a = \frac{e^2}{4\pi\epsilon_0 m_e} \frac{1}{a_0^2 n^4} = \frac{\alpha^3 m_e c^3}{\hbar n^4} \tag{12.18}$$

Plugging back in to the Larmor formula, we get

$$P = \frac{2}{3} \frac{e^2}{4\pi\epsilon_0} \frac{1}{c^3} \left(\frac{e^2}{4\pi\epsilon_0 m_e} \frac{1}{a_0^2 n^4} \right)^2 \tag{12.19}$$

If the electron radiates power at this rate constantly, then the amount of time it can go on radiating, until it has used up all of its energy E, is

$$\tau = \frac{E}{P} \tag{12.20}$$

and so for the corresponding energy

$$\tau = \frac{e^2}{8\pi\epsilon_0} \frac{1}{a_0 n^2} \left(\frac{2}{3} \frac{e^2}{4\pi\epsilon_0} \frac{1}{c^3} \left(\frac{e^2}{4\pi\epsilon_0 m_e} \frac{1}{a_0^2 n^4} \right)^2 \right)^{-1} \tag{12.21}$$

$$= \frac{3}{4} c^3 \left(\frac{e^2}{4\pi\epsilon_0 m_e} \right)^{-2} a_0 n^6 \tag{12.22}$$

$$= (4.7 \times 10^{-11} \text{seconds}) \quad n^6 \tag{12.23}$$

This time could be very large for large n, but we've said that n can't be smaller than 1, which sets a minimum time for the electron to radiate its energy. For $n = 2$, this actually turns out to be remarkably close to the observed time for the electron to go from the $n = 2$ to the $n = 1$ state: the above estimate gives 3 ns, and the correct time is 1.6 ns.

So we now have the following picture:
- Electrons orbiting in an atom have definite energies determined by physical constants and a *quantum number n*.
- Electrons with quantum number $n > 1$ will radiate. The energy carried off by the photon will allow the energy of the electron to change so that $n_{final} < n_{initial}$.
- This process cannot go on indefinitely: n cannot be less than 1, so eventually the atom will find itself in the lowest energy state, the *ground state*

A more complete description of the atom interacting with an electromagnetic field yield the results that the most likely outcome of a transition of the electron's energy between a state with $n = n_2$ and $n = n_1$ is the emission of a *single* photon with a definite energy

$$\boxed{\Delta E = E_{n_2} - E_{n_1} = \frac{m_e c^2}{2}\alpha^2\left(\frac{1}{n_1^2} - \frac{1}{n_2^2}\right)}$$ (12.24)

which translates to a definite wavelength of light

$$\Delta E = h\nu = \frac{hc}{\lambda}$$ (12.25)

For hydrogen, just these wavelengths are what is observed. (We will actually see this for ourselves in class.) Historically, various kinds of transitions were grouped together into so-called *series*: that is, all transitions where the final state is $n_f = 1$ all have fairly high energies (in the ultraviolet) and were named the "Lyman series". Similarly, there are series named after Balmer ($n_f = 2$), Paschen ($n_f = 3$), Brackett ($n_f = 4$), etc. These are primarily useful in classifying an observed transition, but the real description is just in terms of the initial and final quantum number n.

One additional subtlety of the interaction of photons with atoms is that a photon must have *exactly*[1] the energy difference between the two levels in order to cause a transition. If the energy of the photon is less than the energy difference between the levels, nothing happens, but if it's greater, we can't take *part* of the photon energy and use it to cause the transition to happen, and then put the remaining energy somewhere else. The photon is "all-or-nothing". This is quite different from how an electron interacting with the atom would behave, as we shall see below. Examples of this behavior for photons interacting with a hydrogen atom are shown in Figure 12.1.

12.3 A Quantum Mechanical Description of the Atom

Now I should tell you that the above line of argument, while it does get the correct answer for the hydrogen energy (Equation (13.3)), is not wholly correct, and in particular, the expression for the angular momentum Equation (12.3) is not correct. What *is* correct is that for each energy level denoted by n, there is another *quantum number* ℓ, which lies in the range $0 \leqslant \ell \leqslant n - 1$ such that the orbital angular momentum is

$$L = \sqrt{\ell(\ell + 1)}\,\hbar$$ (12.26)

This expression comes from treating the problem in a fully quantum mechanical manner (which we will not do here) using the Schrödinger equation.

The upshot of this is that the angular momentum is not completely determined by the energy (just as in the gravitational problem): electrons in the same energy state can have different angular momenta. Further, and perhaps surprisingly, $\ell = 0$ corresponds to a perfectly reasonable orbit in the quantum description of an atom and it does *not* mean that the electron has fallen to the center of the atom. (Angular

---

[1] Well, "exactly" is too strong, but it must be very close to the energy difference. We'll look at what causes the energy to differ slightly from exact differences.

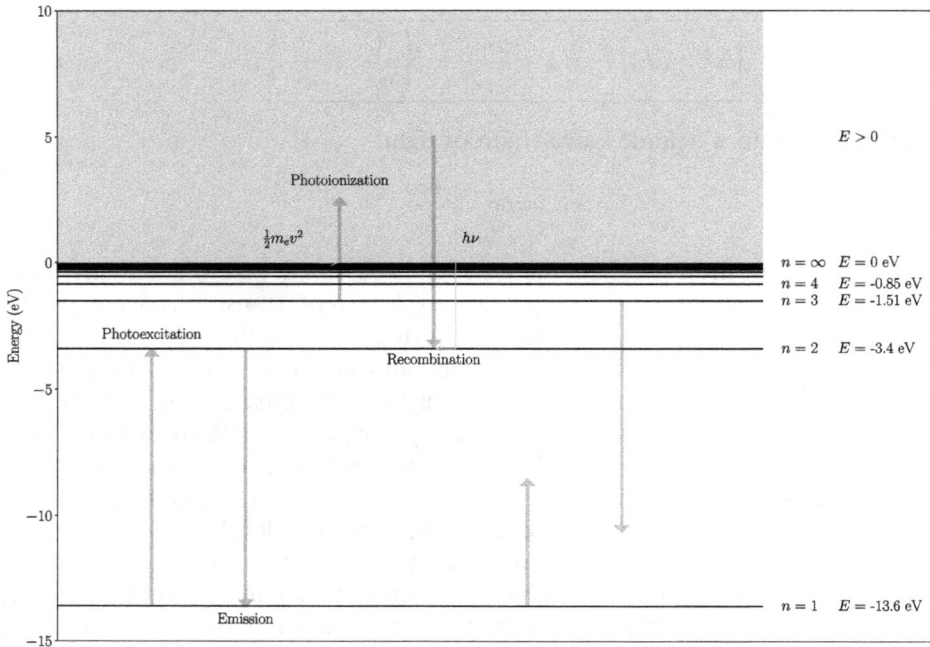

Figure 12.1. An energy level diagram for the hydrogen atom, illustrating several key features of the energy level structure and various processes involving the absorption and emission of photons. The energy levels for the principle quantum number states given by Equation (13.3) are shown as horizontal lines. The quantum number n and energy E_n are given at the right side for the first four levels. Note that, as with the two-body problem, the bound state energy is given as *negative* relative to the energy level of $E = 0$, corresponding to electron that has just been liberated from electrostatic binding to the proton. Note that as $n \to \infty$, the energy levels become increasingly closer spaced, which appears as the thick black band near $E = 0$ in the diagram. Once $E > 0$, the electron is free to assume a continuum of possible states, indicated by the gray region. Several possible energy level transitions of the single electron induced by photons are indicated. Photoexcitation moves the electron from a lower to higher energy state, with the absorption of a photon of exactly the energy difference between states. Emission of photon (whether spontaneous or stimulated) moves the electron from a higher to a lower state, again with the emitted photon having exactly the energy difference. Photoionization promotes an electron from a bound state to one of the continuum, unbound states. The kinetic energy of the resulting freed electron is the difference between the photon energy and the energy of the state: $1/2 m_e v^2 = h\nu + E_n$, where we recall $E_n < 0$. Like photoexcitation, this involves absorbing a photon. Recombination involves a free electron emitting a photon so as to lose energy and end up in a bound state, with the emitted photon having $h\nu = 1/2 m_e v^2 - E_n$. The two transitions indicated in red, where the photon energy does not match the energy different between states *cannot* occur. (In reading this diagram, be aware that the horizontal axis doesn't carry any meaning, the vertical axis only indicates the energy state and should not be thought of as a position in physical space. While the average size of the atom *does* increase as we move upward in the energy diagram, the distance of the electron from the atom is not directly related to the energy once $E > 0$.)

momentum equal to 0 in the gravitational problem would correspond to an ellipse with eccentricity 1.)

To move to a yet more correct description of the atom, we need to revise our ideas about what we call a "state", or equivalently, a physical description of the system.

In the classical two-body problem, specifying E and L (together with the orientation of the ellipse and the location at $t = 0$) allowed us to specify the path taken by the particles at all subsequent times. But in quantum mechanics we find that the path taken by the particle is not well defined—the radius we've been discussing is the average of a probability density for finding the electron at some distance from the proton—but nevertheless E and L remain useful quantities.

Thus we are being pushed toward a description of the atom which includes certain integers, or *quantum numbers*, of which we currently have two: n, which specifies the energy of the orbit, and ℓ, which specifies the angular momentum. It turns out we need two others: an integer m such that $-\ell \leqslant m \leqslant \ell$, which provides the (quantized) projection of angular momentum along a specified axis, usually chosen to be z, and $S = \hbar s / 2$, where $s = \pm 1$, corresponding to a spin of the electron. This "spin" is sometimes analogized as the electron rotating about its own axis, but as far as we know, the electron actually has no physical extent (it is a point), and there is nothing actually "spinning". It just has an intrinsic amount of angular momentum associated with it that has a magnitude of $\hbar / 2$.

In building up an atom more complicated than hydrogen, the quantitative expression we have for the energy of the orbits is no longer correct, but the notion of the state of the electron determined by the quantum numbers n, ℓ, m, and s is still correct. We need to add one additional fact about the universe: all particles come in two varieties: *fermions*, which have an intrinsic spin which is half-integer, and *bosons*, which have integer intrinsic spin. Electrons, protons, and neutrons are all fermions, and photons are bosons. (There are other elementary particles that fall into these categories, which we will not discuss in this course.) Fermions obey the *Pauli exclusion principle*, which states that no two fermions can occupy a state with *all* quantum numbers the same. Bosons are not subject to any such restriction and can occupy the same state in any numbers whatsoever.

Using the quantum numbers and the Pauli exclusion principle, we can now start to build up a description of an atom with Z electrons (and, to preserve charge neutrality, Z protons). These electrons go into the various energy levels, starting at the lowest (the *ground state*) as shown in Table 12.1. You may recognize this as starting to reproduce the patterns observed in the periodic table of the elements.

Table 12.1. Quantum Numbers for the Hydrogen Atom.

n	l	m	s	Electrons per ℓ	Total Electrons
1	0	0	$\pm 1/2$	2	2
2	0	0	$\pm 1/2$	2	8
	1	$-1, 0, 1$	$\pm 1/2$	6	
3	0	0	$\pm 1/2$	2	18
	1	$-1, 0, 1$	$\pm 1/2$	6	
	2	$-2, -1, 0, 1, 2$	$\pm 1/2$	10	

12.4 Atomic Processes

We consider now the various kinds of processes that can occur in atoms, either between two bound quantum states of energy, or between a bound state and the "continuum", that is, when an electron becomes unbound and can have any energy $E > 0$. Note that for these subatomic processes, there is usually an inverse process that can occur. The table below summarizes some common processes and their inverses.

Photoexcitation (a.k.a. absorption)   **Spontaneous Emission**

$X + h\nu \to X^*$ $X^* \to X + h\nu$

Note that $h\nu$ must be *exactly* the energy difference between X and X^*.

Collisional Excitation   **Collisional De-excitation**

$X + \frac{1}{2}mv^2 \to X^* + \frac{1}{2}mv'^2$ $X^* + \frac{1}{2}mv^2 \to X + \frac{1}{2}mv'^2$

$v' \leqslant v, \frac{1}{2}mv^2 \geqslant \Delta E$ $v' \geqslant v, \frac{1}{2}mv'^2 = \frac{1}{2}mv^2 + \Delta E$

Can be a collision with another charged particle, but is usually with an electron.

Electron energy does *not* have to be exactly $X - X^*$

No photon is involved in these processes.

Photoionization   **Recombination**

$X + h\nu \to X^+ + \frac{1}{2}m_e v^2$ $X^+ + \frac{1}{2}m_e v^2 \to X + h\nu$

Note that photon does not need exact energy, just $h\nu > \chi$.

There are two other processes of interest that do not have such a nice parallel between forward and inverse reactions. These are:

Stimulated Emission. Electrons will naturally tend toward the ground state via spontaneous emission. (Note that if the process of reaching the ground state can involve passing through multiple energy levels, photons will be created for each intermediate step in this "cascade" down to the ground state.) But because photons are bosons, they can be stimulated to make a transition where they all pile into the same quantum state. This process is

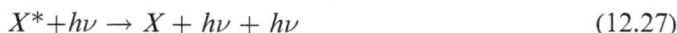

$$X^* + h\nu \to X + h\nu + h\nu \tag{12.27}$$

Collisional ionization Here, the process is similar to collisional excitation, except that the interacting electron has enough energy to actually remove the electron in the atom, and the final state has two free electrons.

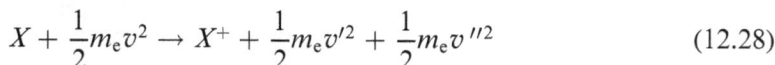

$$X + \frac{1}{2}m_e v^2 \to X^+ + \frac{1}{2}m_e v'^2 + \frac{1}{2}m_e v''^2 \tag{12.28}$$

12.5 Exercise: Atoms and Atomic Processes

In this exercise, we will

- Understand some basic principles of the interaction of light and charged particles with matter

- Gain practice with calculating energies and emitted wavelengths in the hydrogen (and similar) atoms

Question 1. A photon with twice the energy of the $n = 1 \to 2$ transition hits a hydrogen atom, which is in the ground state. Explain what happens to the atom and the photon, and describe the final state of the system after a long time has passed.

Question 2. A photon with half the energy of the $n = 1 \to 2$ transition hits the atom, which is in the ground state. Explain what happens to the atom and the photon, and describe the final state of the system after a long time has passed.

Question 3. An electron with twice the energy of the $n = 1 \to 2$ transition hits the atom, which is in the $n = 2$ state. Describe two possible outcomes of this interaction, again after a long time has passed.

Question 4. An incident electron with a kinetic energy of 12.8 eV strikes a hydrogen atom in its ground state, causing it to change its state. What are the possible wavelengths of photons that are emitted as a result of this interaction? Be sure to think through all the possibilities. It will be helpful to make a list of the changes in electron energies for the various states.

Question 5. What is the maximum speed the incident electron could have after this interaction? Remember, we are assuming the electron causes *some* change in the electron energy in the hydrogen atom (let's ignore the case that the electron has no interaction at all with the hydrogen atom, which is always a possibility).

12.6 Study Questions

1. The following questions all assume photons of wavelength 800 nm (near-infrared light).
    (a) How fast would an electron have to be going to have the same momentum as one photon of this wavelength?
    (b) What number N of such photons would you need to have traveling in the same direction simultaneously to have the same momentum as a person walking across the room (say, 65 kg at 0.5 m s^{-1})?

(c) Suppose that N (from Question 1b) photons per second are being emitted by a source. What is the luminosity (energy per time, in Watts) of that source?

2. Interstellar space is filled with small grains of "dust", made mostly of carbon and silicon atoms. Although the energy levels in a dust grain are different in detail from those of the individual atoms, the idea of energy levels, and of a minimum energy necessary to free an electron from the dust grain, remain the same.

   (a) An ultraviolet photon of wavelength $\lambda = 150$ nm strikes a dust grain and ejects an electron, which leaves the dust grain with a kinetic energy of 3.5 eV. What was the ionization potential from the energy level in which the electron was bound?

   (b) An electron is traveling at 4.5×10^6 m s^{-1}. What is the wavelength of this electron? Compare this wavelength to the typical size of a hydrogen atom.[2]

   (c) Suppose the electron in Item 2b strikes the dust grain in Item 2a and causes an electron to be ejected (similar to what the *photon* did in that question). What would be the speed of this ejected electron, if the electron striking the grain gives up all of its kinetic energy?

   (d) What wavelength would a photon be if it had the same *energy* as the electron in part (Item 2b)? What part of the electromagnetic spectrum would this photon be in?

3. An incident electron with a kinetic energy of 12.8 eV strikes a hydrogen atom in its ground state, causing it to change its state. From the new excited state, the most likely thing to happen is that the atom undergoes (perhaps multiple) spontaneous emission events.

   (a) What are the possible wavelengths of photons that are emitted as a result? Be sure to think through all the possibilities. It will be helpful to make a list of the change in electron energies for the various states.

   (b) What is the maximum speed the incident electron could have after the collision, assuming the electron causes *some* change in the electron energy in the hydrogen atom. (Let's ignore the case that the electron has no interaction at all with the hydrogen atom, which is always a possibility).

---

[2] The electron's velocity is about 1% the speed of light, so it is OK to use the non-relativistic formula for momentum.

An Introduction to Astrophysics with Python
Stars and planets
James Aguirre

Chapter 13

Temperature and Thermodynamics

Having discussed the behavior of individual atoms, we now go on to consider the statistics of large numbers of atoms and introduce the concept of thermal equilibrium and temperature, as well as the notion of a probability distribution. We discuss the Boltzmann factor describing the relative populations of various energy levels in thermal equilibrium, and how, together with the partition function, this allows us to assign to every internal energy state a probability of finding atoms in that state at a given temperature. We also consider what the notion of temperature means for the kinetic energy of particles and introduce the Maxwell–Boltzmann probability distribution for particle energies and velocities.

We have now discussed in great detail the gravitational interaction of two point masses, and how gravitational orbits behave. We've also done a similar thing for individual atoms and how they interact with other atoms, electrons, and photons.

It's now time to add another facet to the discussion of mechanics, basically a *statistical* description of motion for many bodies interacting (gravitationally, electromagnetically, or otherwise) that will allow us to discuss temperature. We have an intuitive sense of what temperature is, but one of the profound realizations of the 19th century was that temperature has relations to *energy*, and also to a measure of *disorder* (entropy). We will not discuss entropy much in this course, but the connection between internal energy states and temperature comes up in many scenarios. Temperature as a concept makes the most sense if we are considering systems in *thermodynamic equilibrium*, which we can operationally take to mean that the rates of every (atomic) process are equal to those of its inverse process: that is, for example, if some fraction of the atoms are being ionized every second, the same number per second are recombining, and for every photon per second emitted of a given frequency, the same number per second are being absorbed. Note that this doesn't mean that no atom is ionized or that there are no photons! What it means is

doi:10.1088/2514-3433/ade5f6ch13
13-1

that there is no net *change* in the fraction of ionized atoms, or the total number of photons of a given frequency. For many practical purposes, we can suppose that thermodynamic equilibrium prevails when the densities of atoms are high enough that collisions or interactions are happening on the same timescale as the various atomic processes.

13.1 Temperature and Energy States

The deep relation between temperature and energy was worked out by Ludwig Boltzmann, who showed that if one has a system, say a gas of interacting particles, with two individual particle energy levels E_i and E_j, with g_i and g_j distinct ways of having that energy, then in equilibrium at a temperature T, the relative numbers n_i and n_j of particles in those states is

$$\frac{n_j}{n_i} = \frac{g_j}{g_i} \exp\left(\frac{-(E_j - E_i)}{kT}\right) \tag{13.1}$$

We will not attempt to derive this relation, but simply make use of it. The *Boltzmann constant* k is there because of a historical accident: temperatures were initially not recognized as a manifestation of energy, and so weren't measured in energy units. So k is the universal constant for converting temperature (a measure of the average energy), which is measured in Kelvin (K), into energy units (J). Thus, we always see factors of kT together. The constant is sometimes written k_B (and available in astropy as c.k_B). Numerically, it is

$$k = 1.38 \times 10^{-23} \text{J K}^{-1} \tag{13.2}$$

We usually order things so that $E_j > E_i$, but this statement is true for any two energy levels i and j.

Our first example for working things out quantitatively is hydrogen, where we have an explicit expression for the energy levels of the neutral atom:

$$E_n = -\frac{m_e c^2}{2} \alpha^2 \frac{1}{n^2} \tag{13.3}$$

We do need to know the *degeneracy* of each energy level, which for H turns out to be

$$g_n = 2n^2 \tag{13.4}$$

Notice that Equation 13.1 doesn't actually give the probability for finding a particle in an energy state E_i, just the *relative numbers* between different states. However, we can turn this into a probability by calculating a quantity called the "partition function" by summing over all the possible states

$$Z = \sum_i g_i \exp\left(-\frac{E_i}{kT}\right) \tag{13.5}$$

Then the probability that an atom has a particular energy E_i is

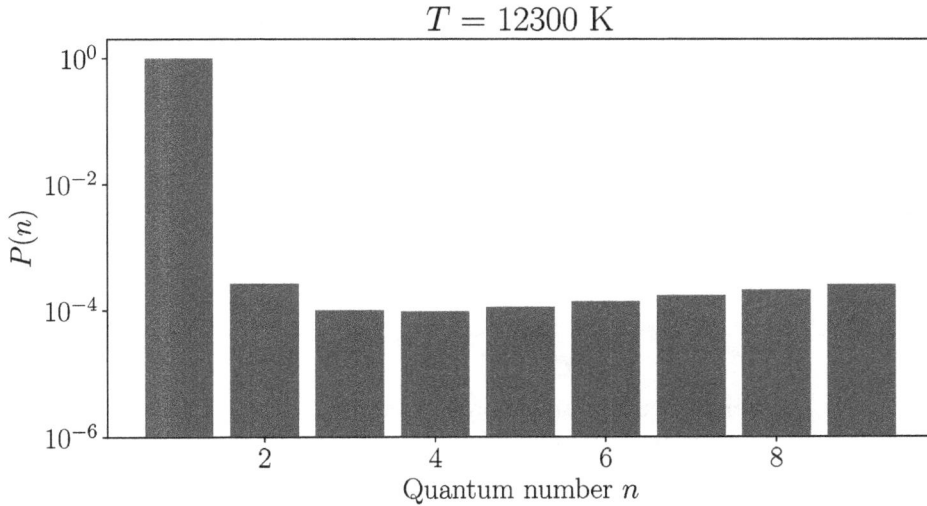

Figure 13.1. The Boltzmann distribution for hydrogen. Here we show the relative probabilities of finding a hydrogen atom in each of the first nine energy levels of hydrogen, assuming the partition function is truncated at $n = 9$, for a temperature of 12,300 K. Note the logarithmic scale on the y-axis to capture the low probabilities of excitation into the $n > 1$ states. This figure is available as an animation, showing the change in probability with temperature (see online at www.doi.org/10.1088/2514-3433/ade5f6).

$$P(E_i) = \frac{1}{Z} g_i \exp\left(-\frac{E_i}{kT}\right) \tag{13.6}$$

We can see that this probability has the desired features that the relative probability of two states is related to the relative size of the populations in each state, as must be true from Equation 13.1:

$$\frac{P(E_j)}{P(E_i)} = \frac{g_j}{g_i} \exp\left(\frac{-(E_j - E_i)}{kT}\right) = \frac{n_j}{n_i} \tag{13.7}$$

and the sum over all states is 1:

$$\sum_i P(E_i) = \sum_i \frac{1}{Z} g_i \exp\left(-\frac{E_i}{kT}\right) = \frac{1}{Z} \sum_i g_i \exp\left(-\frac{E_i}{kT}\right) = \frac{Z}{Z} = 1 \tag{13.8}$$

That is, the sum over all the mutually exclusive possibilities of energy states the atom could be in must imply the atom is in *some* state. An example of a normalized probability distribution for the states of hydrogen is shown in Figure 13.1.

13.2 Temperature and Motion

Now let's consider temperature applied to a perhaps more familiar example, the energy of motion. Specifically, we're going to consider something you've probably seen before, the so-called ideal gas. Here we have a very large number of particles all rushing around, and each has some kinetic energy. If we were to imagine freezing time for an instant, measuring the velocities of all the molecules, and adding up all their kinetic energies, we would get

$$E_{\text{tot}} = \sum_{i=1}^{N} \frac{1}{2} m v_i^2 \qquad (13.9)$$

where we're going to assume there isn't any contribution to the total energy from other sources (potential energy from interactions, or internal energies, for example). Dividing by the total number of particles, we have an average energy

$$\langle E \rangle = \frac{1}{N} \sum_{i=1}^{N} \frac{1}{2} m v_i^2 = \frac{E_{\text{tot}}}{N} \qquad (13.10)$$

Clearly this number is positive, and though at any other instance we would get a different set of v_i as the molecules collide with one another and jostle about, we'd expect that—on average—we'd always get about the same number, assuming I don't add some other source of energy to gas, or let energy out.

Now, what you may not have seen before is how one expects the v_i to be distributed. That is, while we expect the *average* over v_i^2 to be related to the temperature, if I were to make a histogram of the various values of v_i, what would that look like? To answer this, we can apply the ideas of the Boltzmann equation to the motion of particles, where the energy states are not discrete, but continuous, and given by the usual expression

$$E = \frac{1}{2} m \left(v_x^2 + v_y^2 + v_z^2 \right) = \frac{1}{2} m v^2 \qquad (13.11)$$

for each particle. If we consider each each spatial dimension separately, the Boltzmann equation leads us to write the probability for an atom in a gas at temperature T to have a velocity v_z along the z-direction as

$$p(v_z) \propto \exp\left(\frac{\frac{1}{2} m v_z^2}{kT} \right) \qquad (13.12)$$

since the degeneracy of any motion state is the same, and this is just the kinetic energy in the z-direction divided by kT. Note that v_z could be positive or negative. To get a properly normalized probability distribution, it will be helpful to define a quantity with dimensions of velocity:

$$\sigma_v = \sqrt{\frac{kT}{m}} \qquad (13.13)$$

and m is the mass of the atom or molecule. It's customary to write the "molecular" mass in terms of the proton mass, so that

$$m = \mu m_p \qquad (13.14)$$

where μ is a dimensionless quantity, so that

$$\sigma_v = \sqrt{\frac{kT}{\mu m_{\mathrm{p}}}} \tag{13.15}$$

This leads to the probability distribution for *velocities in one dimension* of

$$p(v_z) = \left(\frac{1}{2\pi\sigma_v^2}\right)^{1/2} \exp\left(-\frac{v_z^2}{2\sigma_v^2}\right) \tag{13.16}$$

Some of you may recognize this as the Gaussian distribution for the variable v_z. The most likely velocity (where the function peaks) is zero, as is the average velocity, defined as

$$\langle v \rangle = \int_{-\infty}^{\infty} v_z p(v_z) dv_z \tag{13.17}$$

However, the root-mean-square velocity, defined as

$$\langle v^2 \rangle = \int_{-\infty}^{\infty} v_z^2 p(v_z) dv_z \tag{13.18}$$

is not zero, and is in fact σ_v^2.

We can write the probability distribution for a three-dimensional velocity, assuming all three components of the velocity are independent random variables with the same σ_v, as

$$p(v_x, v_y, v_y) dv_x dv_y dv_z = \left(\frac{1}{2\pi\sigma_v^2}\right)^{1/2} \exp\left(-\frac{v_x^2}{2\sigma_v^2}\right) \times$$

$$\left(\frac{1}{2\pi\sigma_v^2}\right)^{1/2} \exp\left(-\frac{v_y^2}{2\sigma_v^2}\right) \times \tag{13.19}$$

$$\left(\frac{1}{2\pi\sigma_v^2}\right)^{1/2} \exp\left(-\frac{v_z^2}{2\sigma_v^2}\right) dv_x dv_y dv_z$$

$$= \left(\frac{1}{2\pi\sigma_v^2}\right)^{3/2} \exp\left(-\frac{v_x^2 + v_y^2 + v_z^2}{2\sigma_v^2}\right) dv_x dv_y dv_z \tag{13.20}$$

We note that if we transform from a distribution over each Cartesian component separately, to one over just the magnitude, we have the relation

$$p(v_x, v_y, v_z) dv_x dv_y dv_z \rightarrow p(|\mathbf{v}|) dv_x dv_y dv_z \tag{13.21}$$

$$= p(v) 4\pi v^2 dv \tag{13.22}$$

$$= 4\pi \left(\frac{1}{2\pi\sigma_v^2} \right)^{3/2} \exp\left(-\frac{v^2}{2\sigma_v^2} \right) v^2 dv \qquad (13.23)$$

$$= \frac{4\pi}{2\sqrt{2}\,\pi^{3/2}} \frac{1}{\sigma_v^3} \exp\left(-\frac{v^2}{2\sigma_v^2} \right) v^2 dv \qquad (13.24)$$

Simplifying, we have the probability distribution for the *speed in three dimensions*, usually referred to as the Maxwell–Boltzmann distribution:

$$p(v) = \sqrt{\frac{2}{\pi}} \frac{1}{\sigma_v} \left(\frac{v}{\sigma_v} \right)^2 \exp\left(-\frac{v^2}{2\sigma_v^2} \right) \qquad (13.25)$$

The most likely velocity is now no longer zero, but by evaluating the derivative and setting it equal to zero, you can find that it is

$$v_{ml} = \sqrt{2}\,\sigma_v \qquad (13.26)$$

and the average speed is

$$\langle v \rangle = \sqrt{\frac{8}{\pi}}\,\sigma_v \qquad (13.27)$$

Examples of the Maxwell–Boltzmann distribution are shown in Figures 13.2 and 13.3.

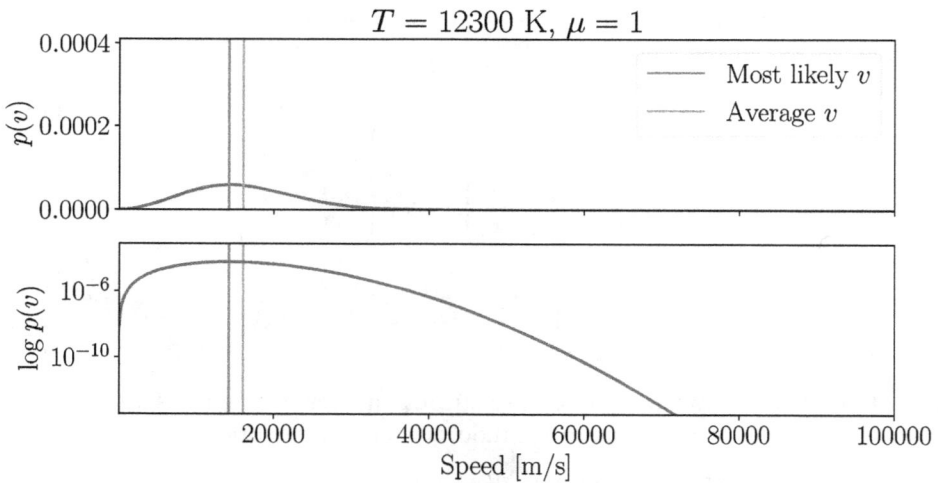

Figure 13.2. The Maxwell–Boltzmann distribution of velocities for hydrogen at a temperature of 12300 K, complementing the picture in Figure 13.1 for the internal energies. Note that the two panels are the same function, but the y-axis is plotted linearly in the top panel to emphasize the peak, and logarithmically in the lower panel to emphasize the tail of the distribution. This figure is available as an animation, showing the change in probability with temperature (see online at www.doi.org/10.1088/2514-3433/ade5f6).

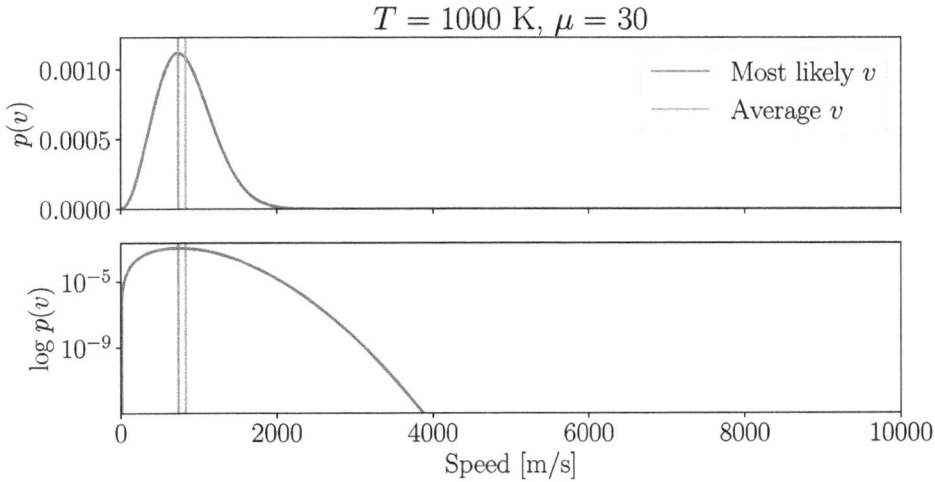

Figure 13.3. The Maxwell–Boltzmann distribution for $\mu = 30$ and $T = 1000$ K, showing a situation relevant to Earth's upper atmosphere. Note that the two panels are the same function, but the y-axis is plotted linearly in the top panel to emphasize the peak, and logarithmically in the lower panel to emphasize the tail of the distribution. This figure is available as an animation, showing the change in probability with μ (see online at www.doi.org/10.1088/2514-3433/ade5f6).

If we now ask about the distribution of *energies*, we can rewrite the probability $p(v)$ to $p(E)$ via a change of variables to get

$$p(E) = \frac{2}{\sqrt{\pi} kT}\left(\frac{E}{kT}\right)^{1/2} \exp\left(-\frac{E}{kT}\right) \tag{13.28}$$

and the average energy is given by

$$\langle E \rangle = \int_0^\infty E p(E) dE = \frac{3}{2} kT \tag{13.29}$$

This lends some precision to the notion that the temperature is in some sense measuring the "average" kinetic energy. Specifically, for an ideal gas

$$T = \frac{2}{3k}\langle E \rangle \tag{13.30}$$

(The only wrinkle is the factor of 2/3.) Also, comparing this to our expression for the *total* energy in Equation (13.10), we see that we could write

$$E_{\text{tot}} = \langle E \rangle N = \frac{3}{2} NkT \tag{13.31}$$

for the ideal gas.

13.3 Exercise: Thermal Escape in Atmospheres

Let's consider a problem of interest for planets: holding on to their atmospheres. We know that if an object achieves escape velocity, it is no longer gravitationally bound. This can happen for molecules of gas in the upper atmosphere. So a sufficiently hot or light molecule might have a thermal velocity high enough to escape the gravitational pull of the planet.

At the end of this exercise, you should

1. Understand the relation between temperature and population of energy states as given by the Boltzmann equation
2. Understand the Maxwell–Boltzmann equation for the distribution of velocities in a gas, and apply this to the escape of gases in an atmosphere

---

Question 1. Calculate the escape velocity from a distance of 500 km above the surface of the Earth. (*Hint*: You've probably written a function to calculate this before. *Another hint*: you don't need to know anything about the type of molecule or the temperature of the atmosphere.)

For the next part, we will need the Maxwell-Boltzmann distribution, and we'll see that it's best if it is done *without* astropy units.

```
@u.quantity_input
def SigmaV(T : u.K,mu) -> (u.m/u.s):
    """ Behavior assumes T is in Kelvin. """
    m = mu*c.m_p
    sigma_v = np.sqrt(c.k_B * T / m)
    return sigma_v

def MaxwellBoltzmann(v, mu, T):
    """ Default behavior is to assume T is in Kelvin an v
    in meters/second, but that these are not astropy Quantity
    obects """
    sigma_v = SigmaV(T*u.K,mu).value
    x = v/sigma_v
    Pv=np.sqrt(2./np.pi)/sigma_v*np.power(x,2)*np.exp(-np
    .power(x,2)/2.)
    return Pv
```

Question 2. Using the Maxwell–Boltzmann function, plot the function $P(v)$ for $T = 1000$ K and $\mu = 28$ (corresponding to N_2 gas) and $\mu = 4$ (for He), for velocities between 0 and 2 times the escape velocity. Use plt.semilogy to better show the range. Label axes! Add vertical lines to show the average velocity of the particles.

We saw, way back in solving differential equations, that we can integrate a function numerically using `scipy.integrate.trapz`. If we have an explicit functional form, we can also use `scipy.integrate.quad`. To integrate our Maxwell–Boltzmann distribution from v_1 to v_2, that is

$$P(v_1 \leqslant v \leqslant v_2) = \int_{v_1}^{v_2} p(v)dv \qquad (13.32)$$

can be done by calling

```
import scipy.integrate as integrate
v1 = 0
v2 = 2000
T = 300
mu = 30
integrate.quad(MaxwellBoltzmann,v1,v2,args=(mu,T))
```

Note that quad wants a function that doesn't have units.

Question 3. Explain why the answer that quad returns in Question 2 makes sense. What is the second number that quad returns?

Question 4. What fraction of the N_2 atoms in the Earth's upper atmosphere $T = 1000$ K are traveling faster than the escape velocity? What about He? What does this suggest about whether Earth will lose either of these gases from its atmosphere?

13.4 Study Questions

1. In addition to what we already know about the Maxwell–Boltzmann distribution for the velocities in a gas, for the following we will need to know the additional fact is that the speed of sound for an ideal gas at temperature T with particles of mass m is

$$v_{\text{sound}} = \sqrt{\frac{1.4kT}{m}} \qquad (13.33)$$

It isn't an accident that this looks similar to our thermal velocities; the collision rate between particles essentially determines how fast information can be transmitted.

(a) Consider a oxygen molecule O_2, the second most abundant in our atmosphere. You can assume $m_{O_2} \approx 32 \ m_{\text{proton}}$.

    i. What is the most likely speed of a oxygen molecule in our atmosphere (assume $T = 300$ K)?

    ii. What is is the average speed of a oxygen molecule in our atmosphere (assume $T = 300$ K)?

    iii. Explain, in words, what your understanding of the difference between "most likely" and "average" is, and why they differ.

    iv. What fraction of the oxygen molecules in 300 K air are moving faster than the speed of sound? You will need to integrate the Maxwell–Boltzmann distribution numerically.

(b) What is the average speed of a hydrogen atom in the atmosphere of the Sun ($T = 5800$ K)? Would a sound wave in the atmosphere of the Sun travel faster or slower than one on Earth?

An Introduction to Astrophysics with Python
Stars and planets
James Aguirre

Chapter 14

Thermal Radiation

Where the previous chapter discussed thermal equilibrium for a gas of massive particles, we now consider what thermal equilibrium looks like for radiation. We begin with some necessary language to describe the radiation field. We then discuss the spectrum of so-called "blackbody radiation" and discuss features of this spectrum. We then conclude with a simple model of the observable surface of a star as a spherical blackbody.

The last chapter considered the probability distribution we should expect for a collection of a very large number of atoms or molecules of average energy approximately kT. We considered two kinds of cases; firstly, that of a large number of hydrogen atoms, and we asked, at a given temperature, what fraction of them we would expect to find with their electrons in different excited states. (This answer was given by the Boltzmann factor, normalized by the partition function, and we saw there were some subtleties with defining this appropriately.) Then we asked for a collection of atoms or molecules mostly with kinetic energy, what the most likely or average speed (or kinetic) would be, or what the probability would be for a molecule to have a speed within some range (with the answer given by the Maxwell–Boltzmann equation). Now we want to consider what it would mean for a collection of *photons* to have a temperature and what the probability distribution would be for their energies. In all these cases, we are forced to make a slightly odd approximation: we need the hydrogen atoms or the molecules or photons to interact with each other, directly or indirectly, so that we can reach a state of thermodynamic equilibrium, but in deriving the formulas, we are writing down the hydrogen atom energies or the molecular kinetic energies or the photon energies as if these particles existed in isolation, without any interactions. The fact that this approximation works pretty well can't really be justified here, but I wanted to point it out, because it's particularly relevant for thermal radiation, where the

doi:10.1088/2514-3433/ade5f6ch14   14-1

photons *don't* interact directly with one another (this is actually a fundamental feature of the electromagnetic field) but rather achieve thermal equilibrium by interacting with matter.

So suppose we have an object with internal energy (and most objects have at least the thermal energy of their constituent atoms), which is sufficiently dense that photons will interact with the matter particles with high probability, each time exchanging energy (equivalent to a photon of a given wavelength), and that these interactions can occur with arbitrary energies. Then, if through these interactions, the photons and matter particles come into *thermodynamic equilibrium*, in which there is no *net* exchange of energy between the particles and radiation, then *the state of this material and the resulting photon spectrum is completely characterized by the temperature T*. We call such an idealized object a *blackbody*, since it is able to interact with and absorb *all* wavelengths of light.

14.1 Describing the Radiation Field

Before we get into the derivation of our blackbody distribution, we're first going to need some vocabulary for describing the behavior of radiation. Though we should think of the final product of our derivation as *like* a probability distribution function, the typical description actually describes not the probability that a photon has a given energy, but rather the distribution of radiated photon energies as a function of wavelength or frequency.

14.1.1 A Brief Digression: Steradians

We're going to find that we need to be able to describe all of the possible different *directions* that the radiation can travel, and although it might seem natural to describe that using a vector, we're going to instead describe the intensity in a given direction using spherical coordinates θ, ϕ to describe the point on the unit sphere toward which the radiation is propagating. Thus, if the intensity traveling toward a point θ, ϕ is described as a function $f(\theta, \phi)$, the total intensity heading into the range of angles $\theta_0 \pm \Delta\theta$, $\phi_0 \pm \Delta\phi$ is given by integrating

$$\int_{\phi_0 - \Delta\phi}^{\phi_0 + \Delta\phi} \left(\int_{\theta_0 - \Delta\theta}^{\theta_0 + \Delta\theta} f(\theta, \phi)\sin\theta d\theta \right) d\phi \qquad (14.1)$$

where

$$d\Omega = \sin\theta d\theta d\phi \qquad (14.2)$$

is a differential "surface area" on the unit sphere.

Recall that if we integrate around the *unit circle* we get 2π radians:

$$\int_0^{2\pi} d\phi = 2\pi \qquad (14.3)$$

Similarly, if we think about integrating over the surface of the unit sphere and write down the angular part of this integral in spherical coordinates, we have

$$\int_0^{2\pi} \int_0^{\pi} \sin\theta d\theta d\phi \equiv \int d\Omega = 4\pi \tag{14.4}$$

The angular integration variable Ω is termed the solid angle, and the steradian is the unit of solid angle. You can think of it as a kind "angular area" (if you think of the radian as "angular length"), and there are 4π steradians in a sphere in the same way that there are 2π radians in a circle.

14.1.2 Definition of the Specific Intensity

We can now describe the flow of electromagnetic energy through a differential surface area dA into a differential range of *solid angle* $d\Omega$ in a differential time interval dt per differential change in frequency $d\nu$ or wavelength $d\lambda$ as

$$dE = I_\nu d\nu dt dA d\Omega \tag{14.5}$$

$$dE = I_\lambda d\lambda dt dA d\Omega \tag{14.6}$$

The quantities I_ν and I_λ have *different* units

$$[I_\nu] = \frac{W}{m^2 \text{ster Hz}} \tag{14.7}$$

and

$$[I_\lambda] = \frac{W}{m^2 \text{ster m}} \tag{14.8}$$

Note that I've written I_λ in such a way as to emphasize that three factors of length should be interpreted as "per unit area, per unit wavelength", rather than as "per volume". If we integrate over wavelength or frequency, thereby considering the total flux per solid angle over the entire electromagnetic spectrum, we get

$$I = \int_0^\infty I_\nu d\nu = \int_0^\infty I_\lambda d\lambda \quad \left[\frac{W}{m^2 \text{ster}}\right] \tag{14.9}$$

Since the total power flux per solid angle I can't depend on the units I use to measure the energy of the light (wavelength or frequency), it must be true that the two versions of the specific intensity are related by

$$I_\nu d\nu = I_\lambda d\lambda \tag{14.10}$$

Notice that you shouldn't ignore the $d\nu$ and $d\lambda$: those are absolutely essential for getting the units to match on both sides; look back at Equations (14.5) and (14.6). This is also expressing the same idea as Equation 14.9, except that it asserts this relation is true for any small range of wavelength or frequency. We'll return to what this means for the specific functional forms later. The quantity I_ν or I_λ is referred to as the *specific intensity*.

14.1.3 Mean Intensity

We can define a variety of new functions that describe some aspect of the radiation field by integrating over the specific intensity in various ways. For example, we can define the specific intensity averaged over all angles as

$$J_\nu = \frac{1}{4\pi} \int I_\nu d\Omega \quad \left[\frac{\text{W}}{\text{m}^2\,\text{Hz ster}}\right] \qquad (14.11)$$

This quantity is usually referred to as the *mean intensity*, and the units are indicated in the brackets following the equation. Note that the units are the same as specific intensity, because we think of the 4π as carrying units of steradians. There is, of course, an equivalent quantity (with different units) if we averaged I_λ over solid angle:

$$J_\lambda = \frac{1}{4\pi} \int I_\lambda d\Omega \quad \left[\frac{\text{W}}{\text{m}^2\text{m ster}}\right] \qquad (14.12)$$

14.1.4 Flux through a Surface

Let's now consider the flux of radiation passing through a surface. We'll write the direction of the radiation flow as $\hat{\mathbf{u}}$, a unit vector directed along the direction of the propagation of the electromagnetic radiation. The differential surface area also has a direction associated with it, namely the normal to the surface at that point. Then we can write

$$L_\lambda = \int I_\lambda(\theta, \phi)\hat{\mathbf{u}} \cdot d\mathbf{A} d\Omega = \int I_\lambda(\theta, \phi)\cos\theta dA d\Omega \qquad (14.13)$$

where θ is the angle between the radiation propagation and the surface normal.

If $I_\nu(\theta, \phi)$ is *isotropic*, that is, it is the same in all directions, or equivalently, it has no angular dependence, we can pull it outside the integral. Then integrating over all directions of the radiation

$$L_\lambda = I_\lambda \int \cos\theta d\Omega dA = I_\lambda \int_A dA \int_{\phi=0}^{2\pi} \int_{\theta=0}^{\pi} \cos\theta \sin\theta d\theta \, d\phi = 0 \qquad (14.14)$$

because of the θ integral. This shouldn't be too surprising: in an isotropic radiation field, there shouldn't be a *net* flux through the surface, because just as much flux is passing one direction from left to right as is passing from right to left.

If we instead ask how much flux passes in *only one* direction (say at the surface of a body), then we only want to integrate θ from 0 to $\pi/2$, and we get

$$L_\lambda = I_\lambda \int \cos\theta d\Omega dA = I_\lambda \int_A dA \int_{\phi=0}^{2\pi} \int_{\theta=0}^{\pi/2} \cos\theta\sin\theta d\theta \, d\phi$$
$$= 2\pi A I_\lambda \int_{\theta=0}^{\pi/2} \cos\theta\sin\theta d\theta \qquad (14.15)$$

The θ integral is now equal to 1/2, and

$$\boxed{L_\lambda d\lambda = A\pi I_\lambda d\lambda}$$

(14.16)

where A is the *total* surface area of the object and L_λ is referred to as *monochromatic luminosity* [W m^{-1}]. Notice that we have performed the area integral in the above, which leads to the monochromatic luminosity, but if we had left the area integral undone, the quantity $L_\lambda/dA \equiv F_\lambda$ we would have a *flux density* [W m^{-1} m^{-2}].

14.1.5 Bolometric Luminosity

Finally, we can compute the integral over area, angle, and wavelength or frequency to get the bolometric luminosity L in Watts (W).

$$L = \int_0^\infty L_\lambda d\lambda = A\pi \int_0^\infty I_\lambda d\lambda$$

(14.17)

14.2 The Planck Formula

The *Planck formula for blackbody radiation* is:

$$\boxed{B_\nu = \frac{2h\nu^3}{c^2}\left[\exp\left(\frac{h\nu}{kT}\right) - 1\right]^{-1}}$$

(14.18)

We can transform B_ν to B_λ by considering the relation

$$B_\lambda = B_\nu \left|\frac{d\nu}{d\lambda}\right|$$

(14.19)

In the function B_ν we substitute $\nu = c/\lambda$ to change the frequency dependence to wavelength dependence, and then we need

$$\frac{d\nu}{d\lambda} = -\frac{c}{\lambda^2}$$

(14.20)

This leads to

$$\boxed{B_\lambda = \frac{2hc^2}{\lambda^5}\frac{1}{\exp\dfrac{hc}{\lambda kT} - 1}}$$

(14.21)

Some important behavior of B_λ in various limits should be noted.
- $\lambda \to 0$, or, equivalently $hc/\lambda \gg kT$. Then

$$\exp\frac{hc}{\lambda kT} \gg 1 \;\Rightarrow\; \exp\frac{hc}{\lambda kT} - 1 \approx \exp\frac{hc}{\lambda kT}$$

(14.22)

and

$$B_\lambda \approx \frac{2hc^2}{\lambda^5} \exp\left(-\frac{hc}{\lambda kT}\right) \tag{14.23}$$

This limit is referred to as the *Wien tail*, and corresponds to the photons behaving like individual particles. Note that the spectrum is strongly (exponentially) suppressed as the photon energies get larger than kT. Also note that (with careful application of L'Hôpital's rule), $B_\lambda \to 0$ as $\lambda \to 0$.

- $\lambda \to \infty$, or, equivalently $hc/\lambda \ll kT$. Then

$$\exp\frac{hc}{\lambda kT} - 1 \approx 1 + \frac{hc}{\lambda kT} - 1 = \frac{hc}{\lambda kT} \tag{14.24}$$

and

$$B_\lambda = \frac{2hc^2}{\lambda^5}\frac{\lambda kT}{hc} = \frac{2ckT}{\lambda^4} \tag{14.25}$$

This is the classical result referred to as the Rayleigh-Jeans law. Notice that the quantum mechanical Planck's constant has canceled out, signaling that this is indeed a classical, wave-like result. Again, with careful application of L'Hôpital's rule, $B_\lambda \to 0$ as $\lambda \to \infty$.

Since B_λ goes to zero in both limits, it must have a peak in-between, which we can find by setting

$$\frac{dB_\lambda}{d\lambda} = 0 \tag{14.26}$$

We expect that the peak in emission occurs when $hc/\lambda \approx kT$, and indeed the wavelength of maximum emission is

$$\lambda_{\text{max}} = \frac{0.0029}{T} \tag{14.27}$$

where λ is in meters, and T in Kelvin. This now gives a way to measure the temperature of an object we can't actually touch: simply measure the wavelength of the peak of its emission! In practice, it's not quite so easy: the peak in the spectrum is fairly broad, and any real-world measurement has noise in it. So we don't actually measure temperature by finding the peak, but conceptually, this is right: we can get a reasonable estimate for the temperature by looking at the relative strengths of the emission from the blackbody at a range of wavelengths. I should also point out that the exact wavelength at which the Planck function peaks should not be taken as the "color" of the object, for two reasons. First, because of the aforementioned breadth of the peak; many wavelengths are typically present, and the peak is not strongly special. Second, the wavelength or frequency at which the Planck function peaks depends on which form of the Equation (A.22) or (A.25) one uses, so clearly is not an intrinsic property of the spectrum. The average energy of a photon is the same in all three cases; see Equation 14.29.

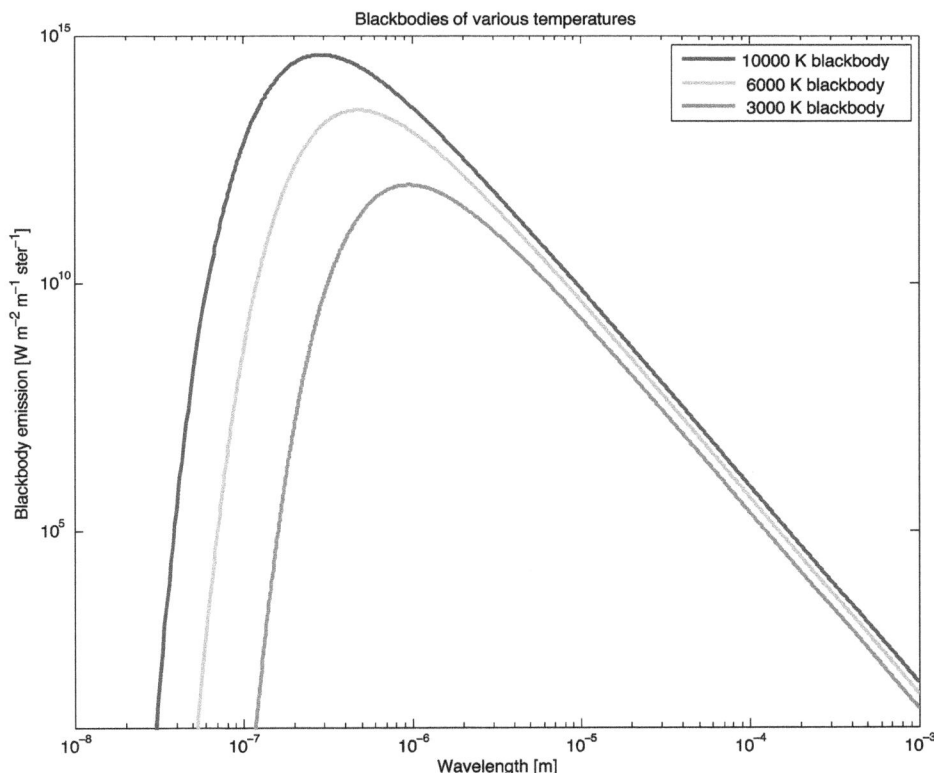

Figure 14.1. Blackbodies at various temperatures. Note that, as expected, the area under the curve (the total power radiated) and that the specific intensity at every wavelength, and the peak wavelength all increase with temperature.

The blackbody function has two other important properties:

1. As T increases, the average energy of emitted photons increases in proportion. (Equivalently, ν_{peak} increases or λ_{peak} decreases.) We say that the emitted spectrum becomes "bluer".

2. If $T_1 > T_2$, then $B_\lambda(T_1) > B_\lambda(T_2)$ for all λ. In other words, increasing the temperature not only increases the peak wavelength, but also the *total* power emitted by the blackbody (i.e., the area under the curve).

It is also true that if $T_2 > T_2$, then $B_\nu(T_1) > B_\nu(T_2)$ for all ν. These properties are illustrated in Figure 14.1.

14.3 Average Energy per Photon and Photon Flux

We can write the *photon flux* as

$$N_{phot} = \pi \int_0^\infty \frac{B_\nu}{h\nu} d\nu = (1.525 \times 10^{15} \text{photons m}^{-2}\text{s}^{-1}\text{K}^{-3}) T^3 \qquad (14.28)$$

The average energy of the photons is given by

$$\langle E_{\text{phot}} \rangle = \frac{\int B_\nu(T)d\nu}{\int B_\nu(T)/(h\nu)d\nu} = (3.729 \times 10^{-23}\,\text{J K}^{-1})T \tag{14.29}$$

14.3.1 Bolometric Luminosity and the Stefan–Boltzmann Law

Using the above definitions, we can calculate the bolometric luminosity (in W) for a blackbody

$$L = A\pi \int_0^\infty B_\lambda d\lambda \tag{14.30}$$

We can show that

$$\int_0^\infty B_\lambda d\lambda = \frac{\sigma T^4}{\pi} \tag{14.31}$$

where σ is the Stefan–Boltzmann constant

$$\sigma = 5.670 \times 10^{-8}\ \frac{\text{W}}{\text{m}^2\text{K}^4} \tag{14.32}$$

(which can be given in terms of fundamental constants). This gives us the Stefan–Boltzmann Law

$$\boxed{L = A\sigma T^4} \tag{14.33}$$

which for a spherical object (like a star!) is $L = 4\pi R^2 \sigma T^4$.

14.4 A Simple Stellar Model

Photons from blackbodies are observed from the points where each photon last interacted with particles, i.e., its *surface*. For stars, which are blackbodies to a good approximation, the photons that we see originate from a thin layer at the surface, the *atmosphere*. This atmosphere is generally not as dense as the layers underneath that produce the true continuum blackbody radiation, so it tends to imprint the signature of the atoms, ions, and, in lower-temperature stars, molecules that are present there.

Nevertheless, the blackbody function now gives us a way to construct our simplest model of the light emission from a star. The star is hot and dense, with a surface temperature T. It is spherical, with radius R, so we have

$$L_\lambda = 4\pi^2 R^2 B_\lambda(T) \tag{14.34}$$

Let's compare the emission spectrum of our Sun to a blackbody; see the following figures. The Sun's peak emission wavelength is $\lambda_{\max} \approx 500$ nm, leading to a surface temperature $T \approx 5800$ K.

As you can see, it's not a perfect match, but it is pretty close. The differences are quite important in what they can tell us, and we will return to them in detail.

The luminosity L is the amount of energy emitted by an object per unit time, and is an intrinsic property of the source (much like mass or radius). However, we do not directly measure the luminosity of a source, but rather the amount of power per unit area from the source reaching us. This quantity is referred to as the *flux*. If all of the energy per time from the source passes through a large sphere centered on the source at radius r, then the flux at this distance will be

$$\boxed{F = \frac{L}{4\pi r^2}}$$

(14.35)

which obviously has units of W m^{-2}.

For the Sun, using a surface temperature of $T \approx 5800$ K and a radius of $R = 6.96 \times 10^8$ m, we can calculate the bolometric luminosity L as

$$L_\odot = 4\pi R^2 \sigma T^4 = 3.91 \times 10^{26} \text{ W}$$

(14.36)

The actual value is

$$L_\odot = 3.84 \times 10^{26} \text{ W}$$

(14.37)

The flux we measure on Earth (above the atmosphere) is

$$F = \frac{L_\odot}{4\pi(1.5 \times 10^{11} \text{ m})^2} = 1365 \frac{\text{W}}{\text{m}^2}$$

(14.38)

which is referred to as the *solar irradiance*.

This simple model of the Sun is compared to the real solar spectrum in Figure 14.2, which notes where the key differences occur. Figure 14.3 shows the location of the hydrogen Balmer ($n_{\text{final}} = 2$) absorption lines, and Figure 14.4 shows the spectrum at short wavelengths, where the differences between the real spectrum and the blackbody approximation are most pronounced, due to absorption lines from a variety of species, termed "line blanketing".

14.5 Exercise: Thermal Radiation

In this exercise, we'll explore some key features of thermal radiation, taking as our example the radiation from our own Sun. Excellent data on the actual spectrum of our Sun can be found from the National Renewable Energy Laboratory, where the data in this exercise was taken from: https://www.nrel.gov/grid/solar-resource/spectra.html

It is necessary to convert the NREL values to a specific intensity for the Sun in SI units.

In this exercise, you will
- Gain experience making sense of a data set and determining appropriate ranges to plot the data
- Understand some features of the blackbody spectrum and its relation to the actual solar spectrum

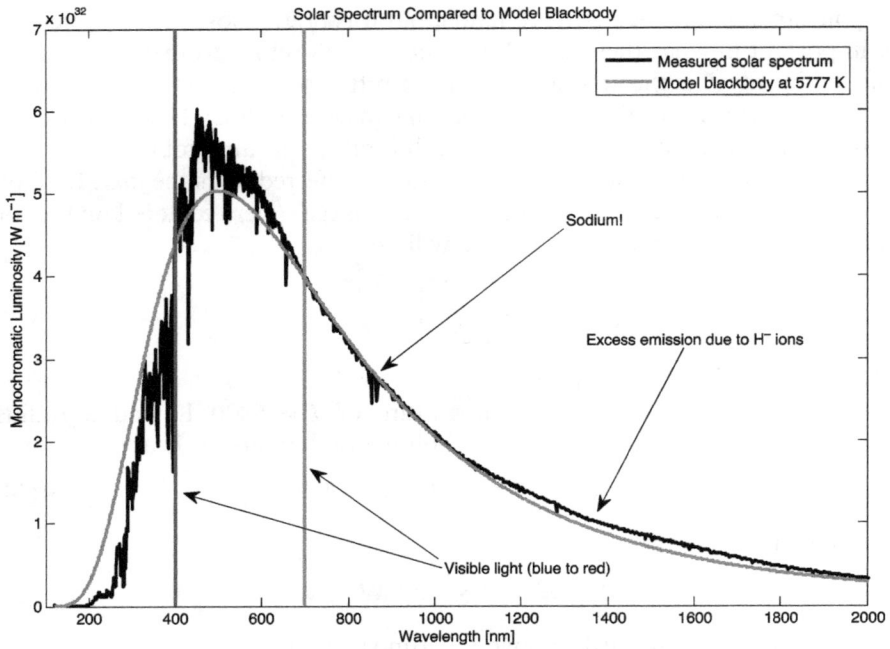

Figure 14.2. The solar reference spectrum compared to a blackbody with an effective temperature to match the bolometric luminosity, with key points of difference noted due to the H$^-$ ion and line blanketing. The range of the spectrum in the visible wavelength range is also indicated.

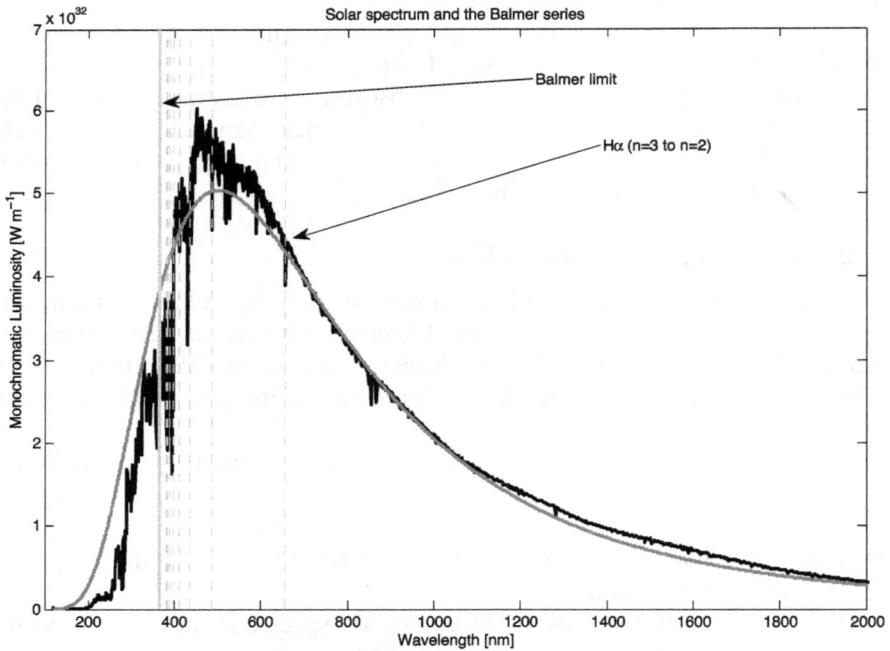

Figure 14.3. The solar reference spectrum, with the Balmer series and Balmer limit indicated.

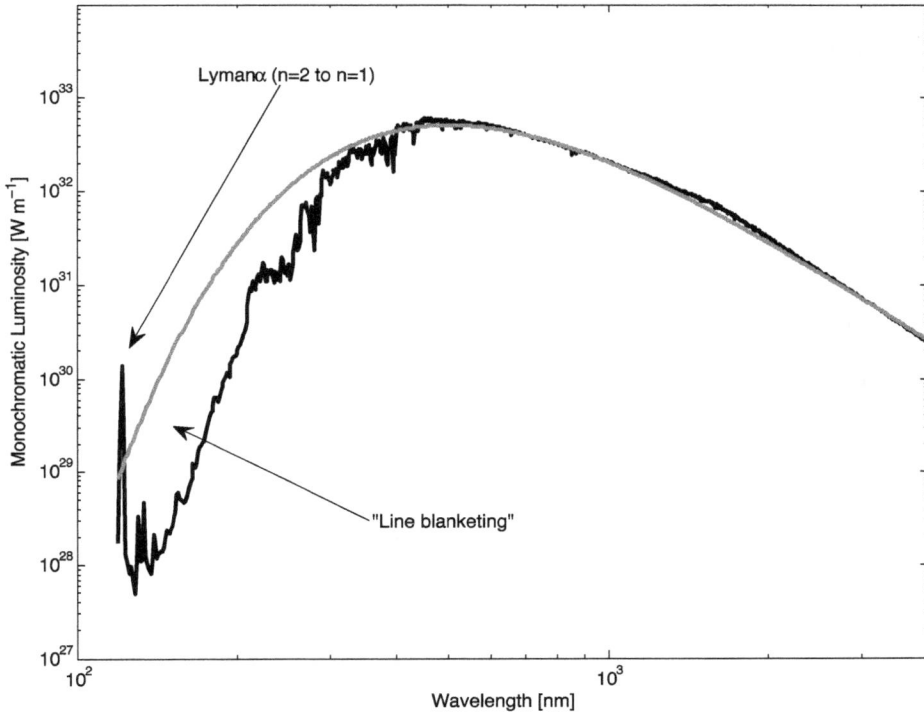

Figure 14.4. A zoom-in on the solar spectrum at short wavelengths, showing the effects of line blanketing near 400 nm, and the Lyman-α emission line.

Question 1. The solar reference spectrum has wavelengths λ given in meters, and specific intensity I_λ with units of W m^{-2} m^{-1}. What is the range of wavelengths provided for the spectrum? Based on the temperature of the Sun, in what range of wavelengths do you expect most of the specific intensity to lie?

Make a plot of the specific intensity versus wavelength, restricting the x-axis to the range of wavelengths where the spectrum is large. Label axes.

Question 2. The total luminosity from the surface of the Sun can be computed by

$$L = \int d\Omega \int dA \int I_\lambda d\lambda = 4\pi R_\odot^2 \pi \int I_\lambda d\lambda \qquad (14.39)$$

Since we only have the solar spectrum as a list of values, we can't use `quad` to integrate it and must use `trapezoid`. Calculate the total luminosity of the Sun (in W) from the above relation. `trapezoid` will correctly handle units. Remember that the units of Ω are steradians. Remember that the radius of the Sun can be gotten from `c.R_sun`, and compare your answer to the constant `c.L_sun`.

14-11

Question 3. Use the relation

$$L = 4\pi R_\odot^2 \sigma_{SB} T^4 \qquad (14.40)$$

to find the surface temperature T of the Sun. The Stefan–Boltzmann constant σ_{SB} is available in astropy as c.sigma_sb.

```
B_lambda_units = 1*u.W/u.m**2/u.sr/u.m
B_nu_units = 1*u.W/u.m**2/u.sr/u.Hz

@u.quantity_input
def BB_nu(T : u.K, wavelength_or_frequency) -> u.W/u.m**2/u
    .sr/u.Hz:

    BB = BlackBody(temperature=T, scale = B_nu_units)

    return BB(wavelength_or_frequency)

@u.quantity_input
def BB_lambda(T : u.K, wavelength_or_frequency) -> u.W/u.m
    **2/u.sr/u.m:

    BB = BlackBody(temperature=T, scale = B_lambda_units)

    return BB(wavelength_or_frequency)
```

Question 4. Use the provided function BB_lambda to plot a blackbody spectrum on top of the actual solar spectrum. How would you describe the differences in the spectra? Find at least one good example of absorption lines in the spectrum, and make a zoomed in plot there.

14.6 Study Questions

1. For this problem, you will need to have functions that return B_λ and B_ν for the Planck function. You can use the ones defined above, or you can write your own. You will be doing numerical integration in this problem and will use both scipy.integrate.quad and scipy.integrate.trapezoid. Note that quad needs a defined function that *does not* use units (you will need to make sure the units are correct yourself) with either ν or λ as the first variable and T as the second. You also need to specify the limits of integration. You can use np.inf for positive infinity, but for this problem I

recommend simply using values that cover the peak of the emission and not going to 0 or ∞ because of numerical problems with the Planck function. Look back at the Exercise in Chapter 13 for how to use quad. trapezoid takes previously defined arrays of numbers for the integrand and the variable of integration, and it is fine with both of those arrays being defined with units and will correctly handle the units in the final answer. For trapezoid you will need to use a list of numbers that sample λ or ν fairly finely and cover the values where B_λ and B_ν are largest.

Note that the best way to specify an array with units is a number array times a quantity object: myarray * u.unit where myarray is a numpy array without units. The syntax [a0 * u.unit, a1 * u.unit, …] where the a_n is a list of quantity objects, tends to cause problems.

(a) Calculate the integral of $B_\lambda(T)d\lambda$ over all λ and $B_\nu(T)d\nu$ over all ν for $T = 5800$ K using both quad and trapezoid. Verify that the integrals over both B_λ and B_ν give you the result $\frac{\sigma}{\pi}T^4$ (check your units, too!) for both quad and trapezoid.

(b) Modify your limits of integration to calculate the *fraction* of the energy of the Sun (which you can assume is a 5800 K blackbody) comes out in the wavelength range 400–700 nm, compared to the total integrated over all wavelengths. Do this for both integrators and both functions B_ν and B_λ. Do you get the same values here for all of them as well?

An Introduction to Astrophysics with Python
Stars and planets
James Aguirre

Chapter 15

Ionization and the Saha Equation

We add one more thermal effect to our repertoire: the possibility of ionization of atoms or molecules. We emphasize that ionization happens at somewhat lower energies than one might naively expect due to consideration of densities. We specifically consider the range of temperatures over which hydrogen ionizes.

Astrophysically, we're going to be interested in ionization because there are many examples of regions where there are a lot of ionizing photons (usually from stars), which produce an ionized plasma where they strike, or situations in which the temperatures are high enough (or, as we'll see, the densities low enough) that many atoms are ionized. Figure 15.1 shows an example of various ionized atoms in the atmosphere of the Sun, emitting at characteristic wavelengths in the ultraviolet.

Let's consider an ensemble of hydrogen atoms at a temperature T.

We have a simple model of the hydrogen energy levels that includes the bound states of the isolated atom, and we have a statistical way of describing the population of them at a temperate T using the fundamental Boltzmann equation (Equation 13.1).

This gives the relative numbers of atoms in the bound energy levels i and j:

$$\frac{n_i}{n_j} = \frac{g_i}{g_j} \exp\left(\frac{-\left(E_i - E_j\right)}{kT}\right) \tag{15.1}$$

We saw that if we wanted to describe the probability that an atom was in a given state, we needed the partition function (Equation 13.5)

$$Z(T) = \sum_n g_n \exp\left(-\frac{(E_n - E_1)}{kT}\right) \tag{15.2}$$

doi:10.1088/2514-3433/ade5f6ch15

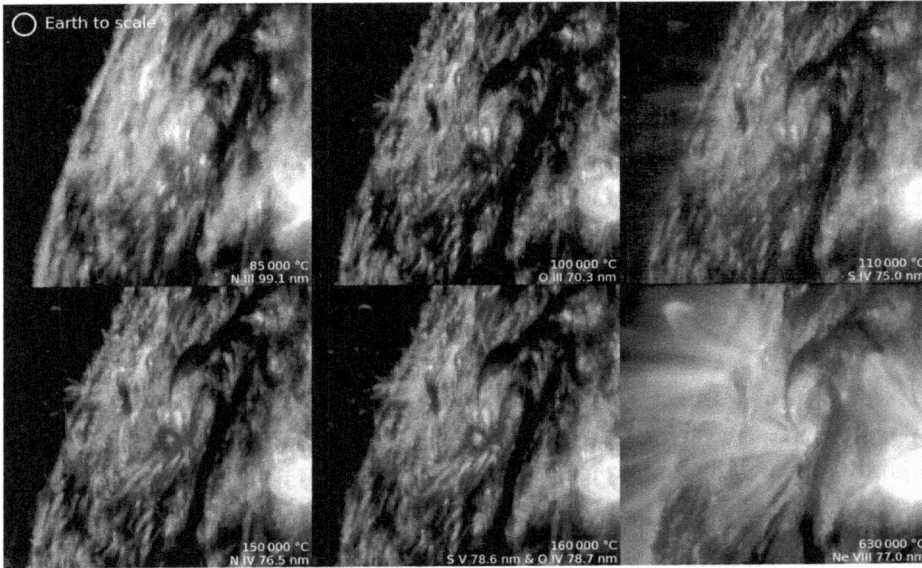

Figure 15.1. A view of the Sun in the light of ionized atoms of nitrogen (N), oxygen (O), sulfur (S), and neon (Ne). The high temperatures in the solar atmosphere produce both ionized atoms as well as atoms and ions with electrons in energy levels above the ground state. Image credit: Image credit: ESA & NASA/Solar Orbiter/SPICE Team https://www.esa.int/var/esa/storage/images/esa_multimedia/images/2022/05/the_sun_s_composition/24071906-1-eng-GB/The_Sun_s_composition_pillars.png

where we have shifted the zero point from zero to the ground state of the hydrogen atom E_1. But notice that the partition function for hydrogen has a problem: if we look at successive terms using the definitions of the degeneracy factor (Equation 13.4) we find that

$$\lim_{n \to \infty} g_n \exp(-(E_n - E_1)/(kT)) = \lim_{n \to \infty} 2n^2 \exp(-(E_n - E_1)/(kT)) = \infty \quad (15.3)$$

Since each term is growing without bound, the sum diverges and so the partition function looks useless. But when we derived the energy levels, we assumed the hydrogen atom was alone in the universe, and the only interaction was between one electron and one proton. The presence of other atoms and charged particles nearby will definitely affect the energy levels. Also, though we did not dwell on it, the average distance of the electron from the proton in hydrogen (Equations 12.6 and 12.7) goes as

$$\langle r \rangle = a_0 n^2 \quad (15.4)$$

so as the principal quantum number increases, the atom becomes larger. Eventually, the distance from the proton to the electron will be less than the average distance to the next hydrogen atom. So something has to give there as well. Although it is not quite right, it turns out that the effect of realistic assumptions about the energy states of hydrogen ends up effectively truncating the series so that it is not, in fact, an infinite sum.

In addition to handling this problem, we need to include not only the bound states, but also the free electron states, with energies described by the Maxwell–Boltzmann distribution

$$p(E) = \frac{2}{\sqrt{\pi}kT}\left(\frac{E}{kT}\right)^{1/2}\exp\left(-\frac{E}{kT}\right) \tag{15.5}$$

which we found in Equation (13.28).

If we correctly account for all the bound and free states of hydrogen (which I will not attempt to do here), we get the *Saha equation*. This gives the relative numbers of ions in state I with Q electrons to an ion $I + 1$ with one *fewer* electrons (i.e., $Q - 1$ electrons):[1]

$$r \equiv \frac{n_{I+1}}{n_I} = \frac{2}{n_e}\frac{Z_{I+1}}{Z_I}\left(\frac{2\pi m_e kT}{h^2}\right)^{3/2}\exp\left(-\frac{\chi_I}{kT}\right) \tag{15.6}$$

The Saha equation has a lot of similarity to the Boltzmann equation (note the ubiquitous exponential function with the ratio of an energy over kT), but there are a lot of things to unpack here. The first is the quantity $Z = Z(T)$, which is the partition function for the *ion* in question. For hydrogen, it turns out that $Z_{\mathrm{HI}} = 2$ is a good approximation for almost all situations (the partition function is basically just that of the $n = 1$ state) and for the proton $Z_{\mathrm{HII}} = 1$ (just one way to be a proton; the proton spin doesn't matter for its behavior as a free particle.) The other quantities of interest in the Saha equation are χ_I, the ionization potential from the ground state of the ion with Q electrons, and n_e, the number density of free electrons.

The ratio r leads to a relatively simple expression to evaluate, but we are usually more interested in the fraction of atoms that are ionized. For hydrogen, where the total number of atoms is simply given by the sum of the ionized and neutral atoms $n_{\mathrm{tot}} = n_{\mathrm{HII}} + n_{\mathrm{HI}}$, little algebra shows that the ionized fraction f is

$$f = \frac{n_{\mathrm{HII}}}{n_{\mathrm{HII}} + n_{\mathrm{HI}}} = \frac{n_{\mathrm{HII}}/n_{\mathrm{HI}}}{n_{\mathrm{HII}}/n_{\mathrm{HI}} + 1} = \frac{r}{1 + r} \tag{15.7}$$

15.1 Temperatures and Ionization

For a gas of particles obeying the Maxwell–Boltzmann distribution, they are all bouncing around, colliding with one another, and the average kinetic energy $\langle K \rangle$ of a particle in a gas at temperature T is

$$\langle K \rangle = \frac{3}{2}kT$$

(referring back to the lecture on Temperature and Thermodynamics). If the energies are sufficiently high, we expect that a single collision might transfer enough energy

---

[1] The reason for the strange notation where $I + 1$ indicates an ion with *fewer* electrons comes from astronomers' odd notation for ions. A neutral carbon atom, for example, would be written Ci (with six electrons), whereas a carbon atom with one electron removed (leaving 5) would be written Cii, and so on, up to Cvi, which has one electron (like a hydrogen atom). Thus we have the common astronomical notation of Hi for neutral hydrogen and Hii for ionized hydrogen (a free electron and proton).

to a neutral atom to ionize it. This would be our naive estimate of the temperature required to ionize hydrogen. The Saha equation is going to give a very different estimate, and it's worth thinking through why that might be. One thing that only focusing on temperature doesn't take into account is the *density* of the gas, and in particular, when an electron is knocked free, how likely it is to find another ionized atom with which to recombine and form a neutral one.

If we want to use the Saha equation to find the relative numbers of Hı and Hıı ions in a gas, unlike the Boltzmann equation, it's clear that we're going to need more than just the temperature and the partition functions (the analog of the degeneracy factors g_i): we also need the electron density n_e. What the Saha equation is telling us is that ionization doesn't just depend on temperature, but also on the density of material, or more precisely, the availability of free electrons.

In many situations, we can infer what the number density of electrons must be by keeping track of all the species of neutral atoms and ions and all of the free electrons, and insisting that no particles leave the volume in question (a so-called "closed box"). We consider a special version of that case below. In other cases, we will assume that the number density of electrons n_e in Equation (7) can be calculated or measured.

15.2 Density and Ionization

For this example of a closed box, consider the atmosphere of a star containing only hydrogen (neutral and ionized). Gravity confines all the particles to an effective constant volume V. In this situation, we can consider the number of free electrons to be equal to the number of Hıı ions:

$$n_e V = N_{\mathrm{HII}} \tag{15.8}$$

Also, the total number of hydrogen atoms (both neutral and ionized), N_t, is related to the density of the gas by

$$N_t = \rho V / (m_p + m_e) \simeq \rho V / m_p \tag{15.9}$$

where m_p is the mass of the proton. (The tiny mass of the electron may safely be ignored in this expression for N_t.) It turns out by making the above substitutions into the Saha equation

$$\frac{N_{\mathrm{HII}}}{N_{\mathrm{HI}}} = \frac{2}{n_e} \frac{Z_{\mathrm{HII}}}{Z_{\mathrm{HI}}} \left(\frac{2\pi m_e kT}{h^2}\right)^{3/2} e^{-\chi_{\mathrm{I}}/kT} \tag{15.10}$$

we can derive a quadratic equation for the fraction of ionized atoms that does *not* depend explicitly on the electron number density. If we set

$$x = \frac{N_{\mathrm{HII}}}{N_t} \tag{15.11}$$

then x solves the equation

$$x^2 + ax - a = 0 \tag{15.12}$$

where

$$a(\rho) = \left(\frac{m_{\mathrm{p}}}{\rho}\right)\left(\frac{2\pi m_e kT}{h^2}\right)^{3/2} e^{-\chi_I/kT} \tag{15.13}$$

Note that while x no longer depends explicitly on n_e, it *does* depend on ρ the *mass density* of the gas, $x = S(\rho, T)$. Thus for this special case, we have an explicit connection between the ionized fraction, the density, and the temperature. Because this is all rather convoluted, the function S is coded up as SahaClosedBoxH.

```python
@u.quantity_input
def SahaRatio(T : u.K, n_e : u.m**-3, Z1 = 2, Z2 = 1, chi =
    13.6*u.eV) -> u.dimensionless_unscaled:

    x = (chi/(c.k_B*T)).to(u.dimensionless_unscaled)
    prefac = 2.*np.pi*c.m_e*c.k_B*T/np.power(c.h,2)
    r = 2 * Z2/(n_e*Z1) * np.power(prefac,1.5) * np.exp(-x)

    return r

@u.quantity_input
def SahaFraction(T : u.K, n_e : u.m**-3, Z1 = 2, Z2 = 1,
    chi = 13.6*u.eV) -> u.dimensionless_unscaled:

    r = SahaRatio(T, n_e, Z1=Z1, Z2=Z2, chi=chi)

    f = r/(1+r)

    return f

@u.quantity_input
def SahaClosedBoxH(T : u.K, rho : u.kg*np.power(u.m,-3)) ->
    u.dimensionless_unscaled :
    """ Calculate the ionized fraction of Hydrogen in a
    closed box of mass density rho and temperature T """

    chi = 13.6*u.eV
    prefac = 2.*np.pi*c.m_e*c.k_B*T/np.power(c.h,2)
    a_coeff = (c.m_p/rho) * np.power(prefac,1.5) * np.exp
(-(chi/(c.k_B*T)).to(u.dimensionless_unscaled))
    x = a_coeff/2. * (np.sqrt(1+4./a_coeff) - 1)

    return x
```

15.3 Exercise: Ionization Zone of a Hydrogen Gas

1. Understand the components of the Saha equation and their meaning
2. Understand the process of thermal ionization described by the Saha equation
3. Calculate the fraction of ionized hydrogen as a function of temperature
4. Understand how ionization is relevant to atoms in the atmospheres of stars

In the following, we'll just consider the ionization of hydrogen, for which we have calculated energy levels that we can use.

Question 1. What is the temperature at which the average particle kinetic energy is just barely sufficient to ionize a hydrogen atom in its ground state?

Keeping that number in mind, now let's examine the various pieces of the Saha equation.

Question 2. First, let's look at the partition function

$$Z(T) = \sum_{j=1}^{\infty} g_j \exp\left(-(E_j - E_1)/(kT)\right) \tag{15.14}$$

where the sum runs over the bound states of hydrogen. Evaluate the first three terms of the sum in Equation (15.14) corresponding to $j = 1$, $j = 2$, $j = 3$ for $T = 1 \times 10^4$ K (chosen to be much lower than the ionization potential). This calculation is greatly simplified by writing a Python function. How good or bad is the approximation $Z \approx 2$ to the case of keeping the first three terms in the partition function sum?

For now, let's take $Z_{HI} = 2$ and $Z_{HII} = 1$ (just one way to be a free proton). The Saha equation for a gas of pure hydrogen (assuming we know n_e) can give us $r = n_{HII}/n_{HI}$.

Question 3. Use a.Saha to calculate r for hydrogen with $n_e = 3 \times 10^{20}$ m^{-3} for a range of temperatures $5000 \leqslant T \leqslant 25,000$ K. (SahaRatio wants all of its inputs to be Quantity objects, that is, to have astropy units.) Plot $r(T)$. Label axes! Describe the behavior of $r(T)$ in words.

Question 4. Use SahaRatio to calculate f_{HII} for hydrogen with $n_e = 3 \times 10^{20}$ m^{-3} for a range of temperatures $5000 \leqslant T \leqslant 25,000$ K. Plot $f_{HII}(T)$. Label axes! Describe its behavior in words. For what temperature is $f \approx 0.5$? $f \approx 0.99$?

Question 5. Going back to the first question, can you think of any reasons why our naive estimate of 10^5 K for the temperature at which hydrogen ionizes is so different from the Saha equation predicts?

One assumption we're making to be able to calculate the fraction of ionized hydrogen at a given temperature is that the number density of electrons n_e is independent of the temperature or the ionization state of the gas. But in reality, we can expect that at low temperatures, n_e is small, whereas at high temperatures it should basically be equal to the number density of protons. We can use the "closed box" approximation to come up with an internally consistent way to derive n_e and give us the ionized fraction $f_{HII} = S(\rho, T)$.

Question 6. Calculate the mass density of the hydrogen gas with $n_e = 3 \times 10^{20}$ m^{-3} when $f_{HII} = 0.5$.

Question 7. Assume the hydrogen gas has the density above, independent of temperature, and use `SahaClosedBoxH` to calculate the ionized fraction as a function of temperature for 5000 K $\leqslant T \leqslant 25,000$ K. (Note that this function returns the ionized fraction, not r like `SahaRatio`.) Plot your result from last time with the fixed value of n_e, and the result from `SahaRatio`. How do they compare generally, and specifically with regard to the points at which $f_{HII} = 0.5$ and $f_{HII} = 0.99$?

Question 8. Now make a plot with f_{HII} for the ρ derived above, as well as 10 times bigger and 10 times smaller. Explain how changing ρ changes the behavior of f_{HII} and physically why this makes sense. (Use `SahaClosedBoxH` for this question.)

Question 9. Calculate the wavelength of the light emitted from transitions between the $n = 2$ and $n = 3$ states in hydrogen.

We call this line H-α or Balmer-α. If we look at a spectrum of our Sun, we see that line in *absorption*. That means there must be some fraction of the total hydrogen atoms that are in the $n = 2$ state at any given time that can absorb photons with the $n = 2 \rightarrow 3$ energy and change to the $n = 3$ state. From this state, the atoms might be collisionally de-excited, might re-emit an Hα photon into some new direction (so it no longer travels to us), or emit from the $n = 3$ back to the $n = 1$ ground state. In any case, some of those $H\alpha$ photons are lost to our view.

So let's answer the question: what fraction of the atoms, which are neutral at a given temperature, are in the $n = 2$ state? First, let's get a sense of whether this fraction is likely to be large or small.

Question 10. Based on your answers above, at the temperature where the gas is 50% ionized, what are the relative numbers of neutral atoms in the $n = 2$ (n_2/n_1) and $n = 3$ (n_3/n_1) states?

To properly find the fraction of atoms in the $n = 2$ state, we need to treat this as the product of two probabilities: the probability that an atom is neutral, and the probability that, if it is neutral, it's in the $n = 2$ state. We can use the Saha equation (in the closed box form) to find the former, and the Boltzmann equation for the latter.

Question 11. Assume it's a reasonable approximation that $n_{HI} = n_1 + n_2 + n_3$ (no atoms are in the $n = 4$ or higher states) for the temperature range $5000 \leqslant T \leqslant 25,000$ K. Derive an expression for $f_2 = n_2/N$. You should write this expression down as an equation. You can code it as a $\texttt{Python}$ function if you like.

Question 12. Make a plot of $f_2(T)$. At what temperature is the largest fraction of atoms in the $n = 2$ state? Approximately what is that fraction?

15.4 Study Questions

1. The Sun is composed primarily of hydrogen. For this problem, we'll ignore the other types of particles in the Sun and assume it is pure hydrogen.[2]
    (a) We've asserted that the problem that the series for $Z(T)$ is divergent (it sums to infinity) is not a real physical problem, because the energy and even the very existence of the higher n energy levels must be modified by the presence of other hydrogen atoms.[3] Let's consider the situation in the solar atmosphere, where we will take the number density of atoms is $n = 6 \times 10^{20}$ m^{-3}.
        i. What is the typical separation between atoms in the solar atmosphere? (Strictly speaking, this question draws from

---

[2] This problem is largely about the limitations of our models; we are going to find that our equations don't always give great results when they're used outside the range of their initial assumptions.

[3] Note that we use n for both number density and principal quantum number in this problem!

material outside of Part II of the course: look at Chapter 17 for the definition of the number density and average particle separation.)

ii. For what value n_{max} of the quantum number n would a hydrogen atom's radius be half the average distance between atoms in the solar atmosphere?

iii. Assume that the partition function sum should be terminated at the n_{max} you found in Item 1(a)ii. At a temperature of $T = 5800$ K, calculate what fraction of the total (neutral) atoms are in the n_{max} state.

(b) In the solar interior, the number density of hydrogen can reach $n = 6 \times 10^{31}$ m^{-3} (which is also the number density of the electrons, as the material is fully ionized) and the temperature is 1.6×10^7 K.

i. Use the Saha equation to find the fraction of hydrogen that is ionized under these conditions. You will *not* find that all the H is ionized. In reality, under these conditions all the hydrogen *is* ionized, so something has gone wrong with our assumptions.

ii. Again, calculate the average spacing between atoms and find n_{max} based on the spacing. How might this fact explain the surprising answer above?

An Introduction to Astrophysics with Python
Stars and planets
James Aguirre

Chapter 16

The Doppler Shift and Its Uses

We present a simple explanation of the Doppler effect and note its importance for measuring velocities. We also show the importance of geometry in this effect, limiting its use to just the line-of-sight velocity. We consider the practical application of this to finding exoplanets and weighing stars.

A very important effect on light occurs when the source of emission of the light is moving. Because of the constancy of the speed of light, the fact that the source is moving means that source's velocity doesn't add to the *velocity* of light. Rather, it changes the perceived wavelength (or frequency), which means the perceived energy as well.

This may be perplexing, since energy is supposed to be a conserved quantity: how is it that I'm changing the energy of a packet of light just having the matter emitting it move? The answer is that we have actually stumbled on a bit of *special relativity*, and to properly answer the question, we need to consider how to describe qualities (like energy or momentum) in frames of reference that are in motion relative to one another. We will not delve into special relativity in this book, but will just attempt to describe why this effect occurs.

If an object is moving at a speed v in a given direction and is emitting light with a wavelength λ, those waves of light travel away from the position of the object at the speed of light c. Let's figure out how those peaks in the wavelength are perceived by a observer that's stationary relative to the moving object by considering a light-emitting object as in Figure 16.1.

We can write the position of the light-emitting object as

$$x = v\, t \tag{16.1}$$

and the radius of the circle of radiation emitted from it at time t_{obs} after it was emitted at t_{emit} is given by

doi:10.1088/2514-3433/ade5f6ch16

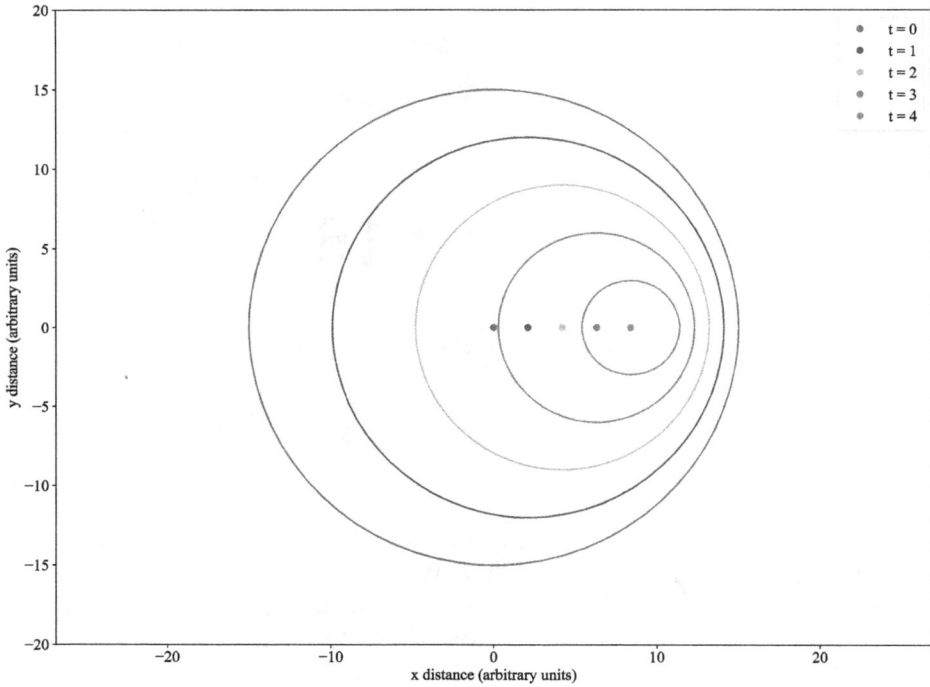

Figure 16.1. The position of an emitter of light (the dot) is shown at different times (shown as different colors) as it moves to the right. The circles show the position of a constant wavefront that started traveling at the time the emitter reached the given position, and is observed at a later fixed time. Note that times and distances are for the observer. In its own frame of reference, the emitter is simply sending out light waves that are always centered on it. An observer sees that the apparent wavelength is shorter (the distance between crests or troughs) in the direction that the emitter is moving, and is longer in the direction the emitter is moving away from.

$$r = c(t_{obs} - t_{emit}) \qquad (16.2)$$

We assume that we have a series of emission times corresponding to a peak in the wave at some wavelength, as seen by the emitter. Then the location of that peak for the observer is a circle centered at the position of the object *when the light was emitted*

$$x_{emit} = v\, t_{emit} \qquad (16.3)$$

This corresponds to an observer getting a snapshot of the wavefronts coming from the emitter at a *fixed* time t_{obs}. (Note that at this fixed time, the position of the emitter itself is not at the location of *any* of the dots; it has already moved on.)

We can notice that the wavelength observed is the distance between successive peaks, and this given by

$$\lambda_{obs} = \lambda_{rest} + v\, \Delta t_{emit} \qquad (16.4)$$

where we have taken the convention that positive v corresponds to motion away from the observer and Δt_{emit} is the distance the emitter travels between successive wave crests. But Δt_{emit} is just given by

16-2

$$\Delta t_{\text{emit}} = \frac{\lambda_{\text{rest}}}{c} \tag{16.5}$$

so we can write

$$\lambda_{\text{obs}} = \lambda_{\text{rest}} \left(1 + \frac{v}{c} \right) \tag{16.6}$$

or in its more familiar form

$$\frac{\lambda_{\text{obs}} - \lambda_{\text{rest}}}{\lambda_{\text{rest}}} \equiv \frac{\Delta\lambda}{\lambda_{\text{rest}}} = \frac{v}{c} \tag{16.7}$$

The above just considered one-dimensional motion. The fully three-dimensional case is given by

$$\frac{\mathbf{v} \cdot \hat{\mathbf{r}}}{c} = \frac{\Delta\lambda}{\lambda_{\text{rest}}} \tag{16.8}$$

where $\mathbf{v}$ is the (relative) velocity between the light source and the observer, and $\hat{\mathbf{r}}$ points along the line between the source and the observer. This is clearer on inspection of Figure 16.2 for the Doppler effect seen by a distant observer. Notice

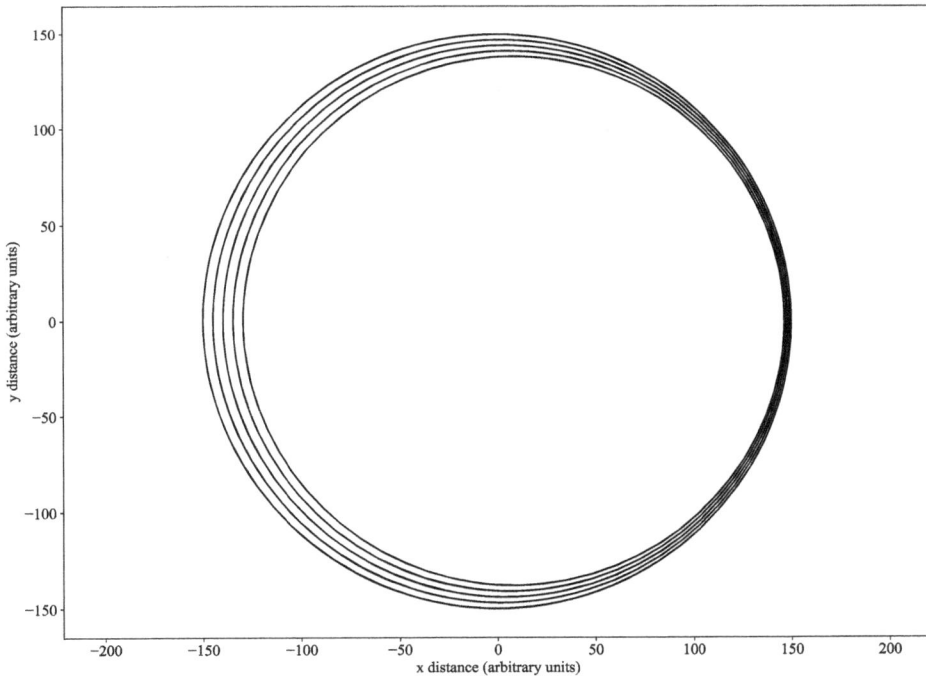

Figure 16.2. Same as the previous figure, but for an observer at a larger distance where the distance the object has moved is much smaller than the distance to the object. In this case, it becomes clearer that the wavelength is unchanged for a far-away observer whose viewing direction is perpendicular to the velocity of the emitter.

that the quantity we actually observe is the wavelength shift (where we've somehow identified the line emitting so that we know λ_{rest}), and then we infer v, either toward or away from us.

As we will see, the Doppler shift is incredibly *useful*: we have seen some indications of how emitted radiation from a body can tell us about its temperature, density, the ionization state of gases associated with it, and its composition. To that list we can now add that light can tell us about the *state of motion* of the material emitting the light.

To take a specific example, consider looking at a planet circling around a star a great distance away. The planet will exert gravitational tugs on the star, causing it to orbit the center of mass of the system. Now we examine an absorption line in the star's atmosphere, known to be at a particular wavelength λ_{rest}. Because of the Doppler shift, we see that wavelength changing slowly with time as the star's velocity is either toward or away from us. Note: this has *nothing* to do with whether the star is actually closer or further away during the motion, but only with how its velocity vector is aligned with our viewing.

Now let's consider the detailed geometry of this situation. To describe the observed Doppler shift in the light from a star with a planet circling around it, we have two effects to consider. One is that the vector $\hat{r}$ to the planet is always very nearly along a Cartesian axis centered on that solar system, say $\hat{x}$, so that we only see the velocity along the *x-axis* for that orbit. The other has to do with the orientation of the plane of the orbit. We call it edge-on if the plane is parallel to $\hat{r}$ (and we would see the planet pass in front of the star) and face-on if the plane is perpendicular to $\hat{r}$ (that is, if we had enough resolution, we would see the planet circling the star). This is shown schematically in Figure 16.3. A particularly striking view of a planetary

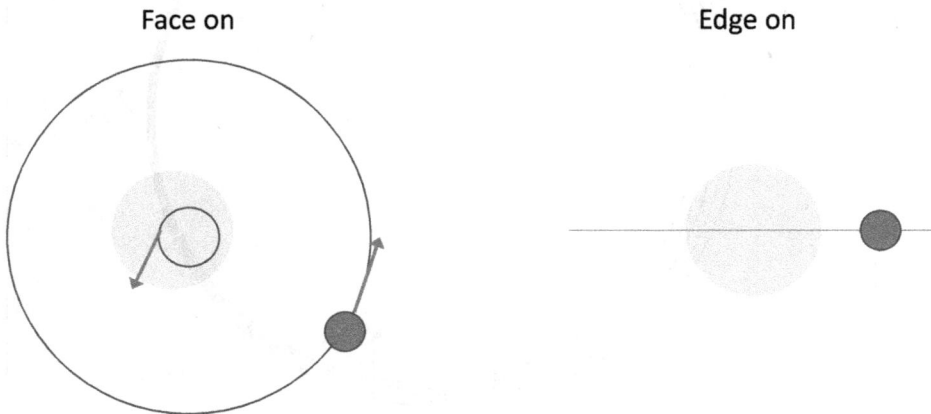

Face on   **Edge on**

Figure 16.3. An illustration of the face-on and edge-on geometries for a planetary system. Note that both the star (yellow) and planet (blue) are orbiting the center of mass, where the orbits are shown in black. Note that both the planet and the star have velocity vectors relative to the viewer. In practice, only the light and the Doppler shift from the star is detectable.

Figure 16.4. ALMA image of the young star HL Tau and its protoplanetary disk. This best image ever of planet formation reveals multiple rings and gaps that herald the presence of emerging planets as they sweep their orbits clear of dust and gas. Line emission from this gas can be used to measure the Doppler shift and rotation velocities, and the effect of the inclination angle of the plane of rotation of the disk. Credit: ALMA (ESO/NAOJ/NRAO); C. Brogan, B. Saxton (NRAO/AUI/NSF). CC BY 4.0

system in formation is shown in Figure 16.4, which clearly shows the system tilted with respect to us.

16.1 Exercise: Finding Exoplanets

In this exercise, we're going to look at a particular application of the Doppler shift, by returning to the two-body problem and considering the motion of star–planet systems where we can observe the motion of the star around the center of mass.

Our goals are to understand the velocity in this simple system and how it manifests as changes in the observed spectrum of the star. The light from the planet is usually impossible to observe directly, as the planet appears so close (in angle) to the star that our telescopes cannot resolve it, and the reflected light from the planet is so much fainter that it is "lost in the glare" of the host star.

Question 1. Consider the Doppler shift of light from a star in a two-body star–planet system. Is the observed wavelength shift due to the Doppler effect larger or smaller when viewing a planetary system nearly edge-on or nearly face-on? Draw a picture explaining why.

Question 2. Plot the x-component of the planet's (object 1) and star's (planet 2) position as a function of time. You will need to put these on two different plots, since the distance from the center of mass is very different for the star and planet. What is the ratio between the maximum distance each gets from the center of mass? When the star reaches its maximum x value, what happens to the planet?

Make a similar pair of plots for v_y for the star and planet. Which has the largest velocity? When the star's velocity is maximum, what is the planet's velocity doing?

Question 3. Suppose that you observe the H-α absorption line, with a rest wavelength of 656.281 nm. Plot the observed *wavelength* of the absorption line as a function of time for the H-α line in the star of the imaginary solar system.

Make a separate plot of the *fractional change in wavelength* as a function of time. How many significant figures in the wavelength do you need to keep to see this fractional change $\Delta\lambda/\lambda$?

Question 4. Explain how you would obtain the semimajor axis of the planet's orbit from measurements of the radial velocity shift of a line in the *star's* atmosphere as a function of time. Assume you can deduce the star's mass from some other method.

Question 5. Now let's consider an eccentric orbit, $e = 0.3$. Compare the plots of v_x and v_y for the star for this case to the $e = 0$ case. Qualitatively, what changes? What is still the same?

16.2 Study Questions

1. In a gas at temperature T, some atoms will be moving toward us and some away based on their distribution of velocities, with a one-dimensional probability distribution (along the line of sight) given an equation in Chapter 13:

$$p(v_z) = \left(\frac{1}{2\pi\sigma_v^2}\right)^{1/2} \exp\left(-\frac{v_z^2}{2\sigma_v^2}\right) \tag{16.9}$$

   Consider the $n = 2 \rightarrow 1$ transition in hydrogen in the Sun's atmosphere. Using the Doppler shift, calculate the wavelength λ_+ of the light emitted by atoms moving at $v_{z,+} = \langle v_z \rangle + \sigma_v$ for and λ_- for ones moving at $v_{z,-} = \langle v_z \rangle - \sigma_v$. What is the fractional spread in wavelengths $(\lambda_+ - \lambda_-)/\lambda_{rest}$?

2. A certain main sequence star has a surface temperature $T = 8200$ K and luminosity of $L = 2$ $L_\odot$ (1 $L_\odot = 3.85 \times 1026$ W). A planet moving around this star at a distance of 2 AU in a circular orbit causes a motion of the star around the center of mass with a maximum velocity of 7.5 m s^{-1}.

   (a) What is the maximum magnitude of the shift in wavelength of the Hα line ($\lambda = 656.28$ nm) that could be observed in this star due to the presence of the planet? Why might the actual observed wavelength shift be smaller than the maximum?

   (b) How many times larger or smaller is the maximum wavelength shift due to the motion of the star as compared to the thermal motion an atom in the star's atmosphere at a velocity given by σ_v in Equation (16.9) Show your calculation. You may want to look ahead to Chapter 22 and the discussion of *thermal Doppler broadening* of lines.

An Introduction to Astrophysics with Python
Stars and planets
James Aguirre

Chapter 17

Introduction to Radiative Transfer

We introduce some concepts necessary to understand radiative transfer, the statistical description of the interaction of radiation with matter. In particular, we discuss the number density and mean spacing of particles in a gas, and the notion of an interaction cross section. We use this to define the optical depth and use this to solve the simplest radiative transfer problem of pure absorption. We build some intuition for the meaning of optical depth.

In developing our theory of atoms and atomic processes, we were really thinking of matter at its most elemental, and we described the interaction of light with matter on the smallest scales, one atom, one photon, and one process at a time. We're going to need a way of talking about the interaction of light and matter in bulk, in some sort of average fashion, where a very large number of interactions are occurring between a very large number of matter particles and photons. This section introduces some of the key ideas needed to describe these interactions, particularly that of photons interacting with matter, or *radiative transfer*. We're building up to the ability to describe how light behaves in, for example, the atmosphere of stars and forms the absorption lines we actually observe.

17.1 Number Density and Mean Free Path

One of the key features of a piece of matter is its *density*. You're no doubt already familiar with this property, defined as

$$\rho = \frac{M}{V} \qquad (17.1)$$

where ρ is the density, M the mass of the object, and V its volume. While this is a useful quantity, often we want to know not the mass per unit volume, but instead the *number* of atoms or molecules or photons[1] per unit volume. This quantity is called the *number density* and is defined as

$$n = \frac{N}{V} \qquad (17.2)$$

where N is the total number of particles. We can describe any material (as solid, liquid, gas, or plasma) by the number densities of the particles that are present. Note that each species of particle can have its own number density in the same volume of space. We'll take up the question of how to account for different species of particles below.

Now, a little thinking will show that knowing the number density also gives us information about how far apart, on average, the particles are. For if the number density is the (average) number per unit volume, then then average volume per unit particle is just the inverse:

$$\mathcal{V} = \frac{1}{n} = \frac{V}{N} = \ell^3 \qquad (17.3)$$

where ℓ is some length that when cubed gives the average volume which contains just one particle. So, remarkably, we can now say the average separation between particles is

$$\ell = \frac{1}{n^{1/3}} \qquad (17.4)$$

If the particles are in a solid, the distance between them is (pretty nearly) fixed, but if we're dealing with liquids, gases, or plasmas, the particles are all moving relative to one another, and thus ℓ is only an average distance between them.

It's worth thinking a little more about what ℓ means for the cases of gases, which we will frequently encounter. If the particles are indeed moving around, then we can expect them to collide occasionally. If you imagine temporarily freezing all particles but one and letting it move, it will typically travel a distance ℓ before colliding with another particle. This leads to the name for ℓ of the *mean free path* (the mean distance over which the particle is traveling freely). How much time passes between collisions? Well, that depends on the particles' velocities, but if a typical velocity is v, then

$$\mathcal{T} = \ell / v \qquad (17.5)$$

is a typical time between collisions. We saw back in Chapter 13 that a large number of (massive) particles in thermal equilibrium will actually have a distribution of velocities described by their temperature and mass. Thus, there is really a distribution of collision times, and we should take expectation of Equation (17.5)

---

[1] In the following, I will use "particle" to mean some sort of object, like an atom, ion, molecule, electron, or photon. The details of what that object is and what it might do are left to a different part of the description.

over this Maxwell–Boltzmann distribution to find a "mean free time" between collisions in this case.

17.2 Cross Section

We now turn to the problem of describing the interactions of particles. We are going to characterize the "strength" or the "likelihood" of particle interactions by a quantity with units of *area* called the "cross-section". At first glance, this may seem an odd choice. To see why this description makes sense, consider the situation shown in Figure 17.1.

Here we have a "slab" of particles of type "a" with number density n_a, and a flux of "b" type particles incident on the slab. We've illustrated the "a" type particles as having some finite projected area, and the "b" type as just arrows. This is because we're often thinking of "a" as some matter particle and "b" as photons. But it's important to understand that the projected area of "a" as seen by the "b" is a *mutual* relation; that is, the cross-section σ_{ab} depends on both types of particle we are considering. (This sounds a bit abstract, but it is really down to the basic physics: an electron, for example, will interact differently with a photon than with an atom, and the cross-section describing these interactions needs to be different as well.)

Having chosen to describe the mutual interaction as an area makes sense if we think of geometry in the figure: there is some total area A through which the "b" particles are passing, and some fraction of it is "blocked" by "a" particles. If we consider very thin slices of thickness dx, then this blocked fraction represents the

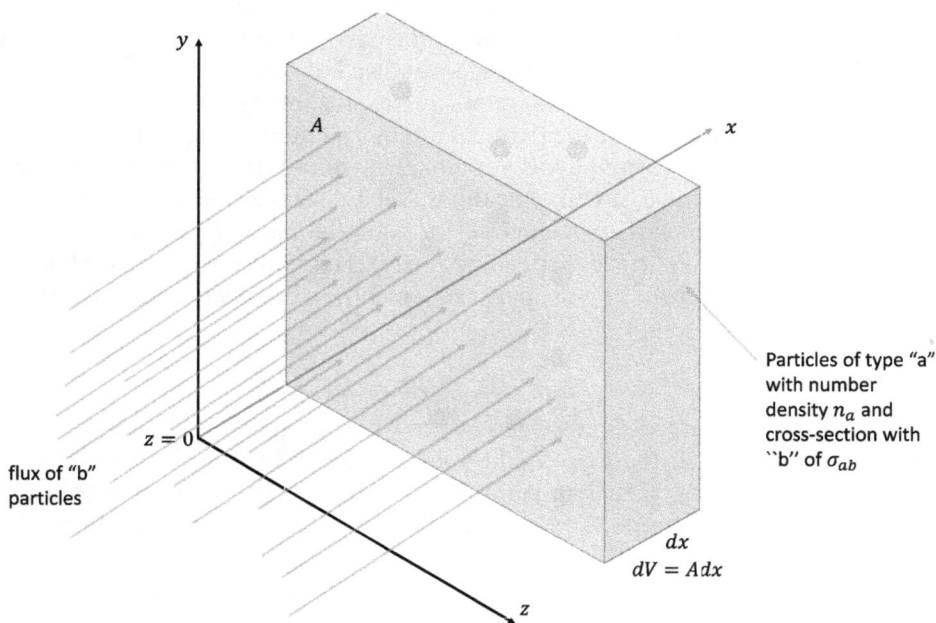

Figure 17.1. A schematic illustration of the quantities involved in defining the cross-section.

change in probability that a "b" particle has interacted with an "a" particle, and we can then integrate to find the total probability of interaction as the "b" particles traverse the slab. So what is the fraction of the area A that is being blocked by "a" particles?

First, we can calculate the total number of particles N_a in the little bit of volume $dV = A dx$ as

$$N_a = n_a dV = n_a A dx \qquad (17.6)$$

Since each of those particles has a projected area σ_{ab}, the total area blocked is

$$A_{\text{blocked}} = \sigma_{ab} N_a = \sigma_{ab} n_a A dx \qquad (17.7)$$

Then the fractional area blocked, which we argue is the small change in probability that a "b" particle interacts with an "a" particle in that volume is

$$d\tau = \frac{A_{\text{blocked}}}{A} = \frac{\sigma_{ab} n_a A dx}{A} = \sigma_{ab} n_a dx \qquad (17.8)$$

where τ is a dimensionless quantity that gives the probability of interaction in a thin slice of material of thickness dx.

All of the above was intended to be quite general, not really caring what the types of particles actually were. Indeed, you may see the concept of cross-sections come up in a lot of other contexts (particle physics, for example). It is worth pointing out that the concepts defined above do rely on some implicit assumptions. In particular, we are assuming that individual interactions between "a" and "b" particles are all independent, i.e., there is no coherent behavior in either group of particles, and we're further assuming that the volume taken up by the "a" particles is a small fraction of the total, in the sense that $\sqrt{\sigma_{ab}} \ll \ell$, and that in-between collisions the particles can be treated as free. All of these assumptions are quite good for the astrophysical gases and plasmas we consider in this book, but it is always good to know the limitations of one's assumptions.

17.3 The Optical Depth

Now, the situation that we will be most interested in is when the "b" particles are photons of a given wavelength λ, and the "a" particles are some particle of matter (an atom or ion, a molecule, a single electron, etc). In this case, we refer to τ as the "optical depth". Then, going back to Equation (17.8), we explicitly write

$$d\tau_{ab}(\lambda) = n_a(\mathbf{x}) \sigma_{ab}(\lambda) dx \qquad (17.9)$$

and

$$\tau(\lambda) = \int_0^x n_a(\mathbf{x}) \sigma_{ab}(\lambda) dx' \qquad (17.10)$$

It is quite common that the number density varies with position, as indicated, though we have assumed the cross-section does not, and is only a function of the wavelength and the type of particles.[2] In the case that the density is constant, we have the simple solution

$$\tau(\lambda) = n_a \sigma_{ab}(\lambda) x \tag{17.11}$$

where x is the total distance the photons traverse.

To see what τ means, let's a consider a flux F (number per unit area per unit time) of incoming photons. Let $F + dF$ be the flux of photons after they traverse a distance dx. Since a fraction $n_a \sigma_{ab} dx$ of the photons are going to interact in that distance, we subtract that bit out, under the assumption that interactions remove the photons from the beam passing through the slab (either by absorbing the photons or by redirecting them so they no longer continue straight through to the other side to be observed). Then the change in flux is the original amount minus the fraction that interacted:

$$F + dF = F - F n_a \sigma_{ab} dx \tag{17.12}$$

Canceling F on both sides and re-arranging, we have

$$\frac{dF}{dx} = -n_a \sigma_{ab} F \tag{17.13}$$

Now, we could proceed to solve this in terms of x, but to relate our answer to τ, let's re-write

$$\frac{dF}{n_a \sigma_{ab}\, dx} \equiv \frac{dF}{d\tau} = -F \tag{17.14}$$

where we have changed variables from x to τ using Equation (17.8). Recalling that $\tau = \tau(\lambda)$, this has the very simple solution

$$F(\tau(\lambda)) = F_0 \exp(-\tau(\lambda)) \tag{17.15}$$

and if we relate τ back to x, we using our constant density assumption Equation (17.11), we have

$$F(x) = F_0 \exp(-n_a \sigma_{ab}(\lambda) x) = F_0 \exp(-x/\ell_{ab}(\lambda)) \tag{17.16}$$

where we have defined a distance ℓ_{ab}, called the *mean free path*[3] for interactions as

$$\ell_{ab}(\lambda) = \frac{1}{n_a \sigma_{ab}(\lambda)} \tag{17.17}$$

---

[2] The cross-section could also vary with position, for example, if it itself depends on the density of the matter particles, but we will not consider such cases in this book.
[3] Notice that though I am also calling this a mean free path, it is *not* the same as that defined in Equation (17.4), which just specified how far apart on average some particles with a given number density are. The length defined here is how far on average a "b" particle will travel before interacting with an "a" particle, and is not just related to the number density of "a" or "b" particles, but includes the cross-section between them as well.

Since Equation (17.16) is measuring the flux that gets through, that means that after traveling a distance ℓ_{ab},

$$F(x = \ell_{ab}) = 0.368 F_0 \qquad (17.18)$$

that is, $1/e = 0.368 = 36.8\%$ of the incident photons are transmitted, i.e., do *not* interact with the "a" particles and the remaining fraction, or 0.723 of the particles, *did* interact and were absorbed.

Another useful quantity is the *mean time between successive collisions*. If we assume the photons travel at the speed of light between collisions, this gives

$$\mathcal{T} = \frac{\ell_{ab}}{c} \qquad (17.19)$$

The *collision frequency* is just

$$\nu = \frac{1}{\mathcal{T}} = \frac{c}{\ell_{ab}} = c n_a \sigma_{ab} \qquad (17.20)$$

17.4 Low and High Optical Depth

We can use the solution to this problem of "pure absorption" given by Equation (17.15) to gain some insight into the meaning of the optical depth. We can see that $\tau = 0$ corresponds to no absorption, and all of the light passes through the slab unabsorbed. By contrast, the limit $\tau \to \infty$ means $\exp(-\tau) \to 0$ and *no* light is transmitted. All of this suggests that, for this case at least, we can interpret $\exp(-\tau(\lambda))$ as the probability that a photon of wavelength λ did *not* interact in crossing the slab, and $1 - \exp(-\tau(\lambda))$ as the probability that it did. These quantities interpreted as probabilities do indeed have the necessary feature that they are always between 0 and 1, unlike τ itself, which can exceed 1 (according to the definition).

Under this interpretation, we can also explore what happens for small $\tau \ll 1$ and large $\tau \gg 1$. In the case $\tau \ll 1$, we can Taylor expand the exponential around 0 to find

$$\exp(-\tau) \approx 1 - \tau \qquad (17.21)$$

and so the probability of absorption is $1 - \exp(-\tau) \approx 1 - (1 - \tau) = \tau$, which is in line with our original derivation which equated the probability of interaction as being equal to τ, in the limit that τ was small.

For insight into larger τ, let's consider one of the more profound and ubiquitous concepts in statistics and physics, the *random walk*. Let's assume that a given photon pings along a material with a mean free path[4] for interaction of ℓ, and each time it interacts it changes direction, but continues to travel, on average, a distance ℓ before the next collision. How far does it travel in total? Let's imagine the motion as a series

---

[4] The attentive reader will note that this should properly be ℓ_{ab}, but I have dropped the subscripts describing the interaction to simplify the notation.

of j steps, described by vectors $\mathbf{l}_j$, representing the motion of the particle between collisions. The average length of each of these steps is the mean free path ℓ, but the actual length and orientation of any step is random. The total distance traveled can just be calculated by the vector sum over all the steps; we take the total number to be N:

$$\mathbf{d} = \sum_{j=1}^{N} \mathbf{l}_j \tag{17.22}$$

To calculate the length of the vector $\mathbf{d}$, we need to calculate its magnitude, or $|vectd|^2 = \mathbf{d} \cdot \mathbf{d}$. We will need to define the dot product between the vectors defining the individual steps as

$$\mathbf{l}_i \cdot \mathbf{l}_j = \ell_i \ell_j \cos(\theta_{ij}) \tag{17.23}$$

where $\ell_{i,j}$ are the lengths of $\mathbf{l}_{i,j}$ and θ_{ij} is the angle between them (in the plane containing both of them). Then we can simply calculate

$$\mathbf{d} \cdot \mathbf{d} = \left(\sum_i \mathbf{l}_i\right) \cdot \left(\sum_j \mathbf{l}_j\right) = \sum_{ij} \ell_i \ell_j \cos(\theta_{ij}) = \sum_i \ell_i^2 + \sum_{i \neq j} \ell_i \ell_j \cos(\theta_{ij}) \tag{17.24}$$

where I have separated the terms where a vector is dotted with itself from those where it is dotted with a distinct vector. Now comes the part where we invoke the randomness of the walk. We're going to argue that averaging[5] over the cosine of random angles gives zero (equal numbers of values both greater and less than zero, since there's no preferred direction):

$$\left\langle \cos(\theta_{ij}) \right\rangle = 0 \tag{17.25}$$

We also argue that the average of the square of a step is the square of the mean free path:

$$\langle \ell_i \ell_i \rangle = \ell^2 \tag{17.26}$$

This lets us argue that the average of the squared distance traveled is

$$\langle \mathbf{d} \cdot \mathbf{d} \rangle = \left\langle \sum_{ij} \ell_i \ell_j \cos(\theta_{ij}) \right\rangle \tag{17.27}$$

$$= \left\langle \sum_{i=1}^{N} \ell_i^2 \right\rangle + \left\langle \sum_{i \neq j} \ell_i \ell_j \cos(\theta_{ij}) \right\rangle \tag{17.28}$$

$$= \sum_{i=1}^{N} \langle \ell_i^2 \rangle + \sum_{i \neq j} \langle \ell_i \ell_j \rangle \left\langle \cos(\theta_{ij}) \right\rangle \tag{17.29}$$

$$= \sum_{i=1}^{N} \ell^2 + \ell^2 \sum_{i \neq j} \left\langle \cos(\theta_{ij}) \right\rangle \tag{17.30}$$

---

[5] There are a lot of subtleties to averaging correctly, but here we're going to use $\langle\rangle$ to mean the average we get in the limit as the number of steps goes to infinity.

$$= N\ell^2 \tag{17.31}$$

Thus we have that

$$\tau = n\sigma\sqrt{N}\ell = \sqrt{N} \tag{17.32}$$

using the definition of the mean free path. Thus, in the limit of a large optical depth, τ measures the square root of the number of times the photon has interacted.

17.5 Exercise: Optical Depth for Free Electrons

In the first part of this exercise, you'll put some numbers to the notions of number density and mean spacing of particles for situations of both terrestrial and astrophysical interest. Then we'll use these ideas to calculate the optical depth of several astrophysically interesting cases.

Question 1. Let's work out the mean spacing between molecules of carbon (atomic mass 12) in the graphite of a pencil "lead". (A pencil is an obsolete device for making marks on a piece of paper.) The mass density of graphite is 2.3 g cm^{-3}.

What is the number density C atoms in graphite in cm^{-3}?

What is the typical spacing between C atoms? Express this in nm (10^{-9} m) and relative to the radius of hydrogen atom (using the constant c.a0).

Question 2. Let's work out the mean spacing between molecules of air in the room, using the ideal gas law $P = nkT$. Let the temperature $T = 300$ K and the pressure to be 101325 Pascal (Pa), typical atmospheric pressure.

What is the number density of air molecules in cm^{-3}?

What is the typical spacing between molecules? Express this in nm (10^{-9} m) and relative to the radius of hydrogen atom (using the constant c.a0).

Question 3. A typical number density for atoms in an interstellar gas cloud might be $n = 300$ cm^{-3}, with the temperature of the cloud being 30 K, which we'll assume is composed completely of hydrogen molecules, H_2.

How does this number density compare to that of molecules in the air of this room or atoms in a pencil lead?

What is the mean time between collisions of H_2 molecules in this cloud? (Recall the relation between particle velocity and temperature.) Is it reasonable to think that collisions occur often enough to keep the atoms in thermal equilibrium, if the lifetime of the cloud is measured in millions of years?

Question 4. The cross-section for the interaction of a photon with an energy much less than $m_e c^2$ with an electron is called the Thompson cross-section $\sigma_T = 6.652 \times 10^{-29} \text{m}^2$, available as `c.sigma_T`. To figure out the number density of electrons, assume that 10^{-6} of the hydrogen molecules in the cloud in Question 3 have been ionized.

What is the number density of free electrons in the cloud?

If the cloud is roughly a sphere 30 light years in diameter, what is the optical depth for photons crossing the thickest part of the cloud, assuming they only interact with the electrons, and not with the molecules?

Question 5. Let's work out the optical depth for scattering off a free electron in the solar atmosphere, assuming a number density of electrons $3 \times 10^{20} \text{ m}^{-3}$ and thickness of 300 km. (We justify this thickness a bit more in Chapter 20.) What number density of electrons would be required in the previous problem so that only 10% of photons were able to get through a thickness of one scale height of 300 km?

An Introduction to Astrophysics with Python
Stars and planets
James Aguirre

Chapter 18

Summary of Light, Matter and Temperature

We summarize key ideas from the middle part of the book, including ideas from thermodynamics and statistical mechanics, quantum mechanics, and the interaction of light and matter.

At this point in the book it is well worth taking a little time to think about the bigger picture of what we are learning, and how it relates to science more broadly. Here are some thoughts:

1. Light is a key carrier of information for astrophysics. It can tell us about the temperature, density, ionization state, composition (what elements are present), and state of motion of the matter which emitted the light.

2. Light behaves as both a particle and a wave. The speed of light is related to the wavelength and frequency of the wave

$$c = \lambda \nu$$

but the particles of light (photons) carry energy in discrete chunks

$$E = h\nu$$

The photons also carry momentum

$$p_{\text{light}} = \frac{h}{\lambda} = \frac{h\nu}{c} = \frac{E}{c}$$

Remarkably, it also turns out that matter behaves both as a particle and a wave, with momentum

$$p_{\text{matter}} = mv = \frac{h}{\lambda}$$

and (kinetic) energy

$$E = \frac{1}{2}mv^2 = \frac{1}{2m}\left(\frac{h}{\lambda}\right)^2$$

doi:10.1088/2514-3433/ade5f6ch18

3. A generic feature of the energy levels of electrons in matter is that they have distinct, discrete energy levels. Our favorite example is hydrogen, for which the energy levels are given by

$$E_n = -\frac{m_e c^2}{2}\alpha^2 \frac{1}{n^2}$$

with n being an integer. Transitions between energy levels can occur only for specific energy differences. For hydrogen, these are

$$\Delta E = E_{n_2} - E_{n_1} = \frac{m_e c^2}{2}\alpha^2\left(\frac{1}{n_1^2} - \frac{1}{n_2^2}\right)$$

Transitions between these levels can either absorb or emit very specific wavelengths of light, called emission or absorption lines

$$\Delta E = h\nu = \frac{hc}{\lambda}$$

For hydrogen, it makes sense to speak of the average distance of the electron from the nucleus

$$\langle r \rangle = \frac{4\pi\epsilon_0}{e^2\mu}\hbar^2 n^2 \equiv a_0 n^2$$

4. Various kinds of processes can occur in atoms either between two bound quantum states of energy, or between a bound state and the "continuum", that is, when an electron becomes unbound and can have any energy $E > 0$. Note that for these subatomic processes, there is usually an inverse process that can occur. The table below summarizes some common processes and their inverses.

Photoexcitation (a.k.a. absorption)
$X + h\nu \rightarrow X^*$

Spontaneous Emission
$X^* \rightarrow X + h\nu$

Note that $h\nu$ must be *exactly* the energy difference between X and X^*.

Collisional Excitation
$X + \frac{1}{2}mv^2 \rightarrow X^* + \frac{1}{2}mv'^2$
$v' \leqslant v, \frac{1}{2}mv^2 \geqslant \Delta E$

Collisional De-excitation
$X^* + \frac{1}{2}mv^2 \rightarrow X + \frac{1}{2}mv'^2$
$v' \geqslant v, \frac{1}{2}mv'^2 = \frac{1}{2}mv^2 + \Delta E$

Can be a collision with another charged particle, but is usually with an electron.
Electron energy does *not* have to be exactly $X - X^*$
No photon is involved in these processes.

Photoionization
$X + h\nu \rightarrow X^+ + \frac{1}{2}m_e v^2$
Note that photon does not need exact energy, just $h\nu > \chi$.

Recombination
$X^+ + \frac{1}{2}m_e v^2 \rightarrow X + h\nu$

There are two other processes of interest that do not have such a nice parallel between forward and inverse reactions. These are: Stimulated Emission. Electrons will naturally tend toward the ground state via spontaneous emission. (Note that if the process of reaching the ground state can involve passing through multiple energy levels, photons will be created for each intermediate step in this "cascade" down to the ground state.) But because photons are bosons, they can be stimulated to make a transition where they all pile into the same quantum state. This process is

$$X* + h\nu \rightarrow X + h\nu + h\nu$$

Collisional ionization Here, the process is similar to collisional excitation, except that the interacting electron has enough energy to actually remove the electron in the atom, and the final state has two free electrons.

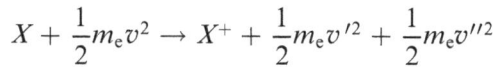

$$X + \frac{1}{2}m_e v^2 \rightarrow X^+ + \frac{1}{2}m_e v'^2 + \frac{1}{2}m_e v''^2$$

5. When discussing the bulk behavior of matter in which a state of *thermodynamic equilibrium* prevails between all processes, it it useful to introduce the notion of temperature. The ratio between the average number of particles in two energy states at a given temperature is given by the *Boltzmann equation*:

$$\frac{n_2}{n_1} = \frac{g_2}{g_1} \exp\left(-\frac{\Delta E}{kT}\right) = \frac{g_2}{g_1} \exp\left(-\frac{(E_2 - E_1)}{kT}\right)$$

The actual probability for a particle to be in a particular energy state is given by

$$P(E_j) = \frac{1}{Z(T)} g_j \exp\left(-\frac{(E_j - E_1)}{kT}\right)$$

where Z is the *partition function*

$$Z(T) = \sum_{n=1}^{\infty} g_n \exp\left(-\frac{(E_n - E_1)}{kT}\right)$$

When applied to the freely moving particles of an *ideal gas*, this leads to a probability distribution for the velocities in any one direction of

$$p(v_z) \propto \exp\left(\frac{\frac{1}{2}mv_z^2}{kT}\right)$$

Defining the *mean molecular weight*

$$\mu = \frac{m}{m_p}$$

and the quantity

$$\sigma_v = \sqrt{\frac{kT}{\mu m_p}}$$

the probability for the full three-dimensional velocity is given by the *Maxwell–Boltzmann distribution* in terms of speed

$$p(v) = \sqrt{\frac{2}{\pi}} \frac{1}{\sigma_v} \left(\frac{v}{\sigma_v}\right)^2 \exp\left(-\frac{v^2}{2\sigma_v^2}\right)$$

or energy

$$p(E) = \frac{2}{\sqrt{\pi}kT} \left(\frac{E}{kT}\right)^{1/2} \exp\left(-\frac{E}{kT}\right)$$

6. Sufficiently dense objects (which have a near-continuum of energy levels) will emit radiation with *Planck spectrum of blackbody radiation*, which can be expressed per unit frequency

$$B_\nu = \frac{2h\nu^3}{c^2} \left[\exp\left(\frac{h\nu}{kT}\right) - 1\right]^{-1}$$

or per unit wavelength

$$B_\lambda = \frac{2hc^2}{\lambda^5} \frac{1}{\exp\dfrac{hc}{\lambda kT} - 1}$$

In either case, the average energy per photon from a thermal emitter with temperature T is given by

$$\langle E_{\text{phot}} \rangle = (3.729 \times 10^{-23} \text{J K}^{-1}) \quad T$$

The *monochromatic luminosity* (power per unit wavelength) of a blackbody of surface area A is

$$L_\lambda d\lambda = A\pi B_\lambda d\lambda$$

and a *bolometric luminosity* (power integrated over all wavelengths) given by the Stefan–Boltzmann Law

$$L = A\sigma T^4$$

7. The relative numbers of an atom or ion denoted I and an ion with one *fewer* electron denoted $I + 1$, where the atoms and ions are in thermal equilibrium with a radiation field, all at temperature T, is described by the *Saha equation*

$$\frac{n_e n_{I+1}}{n_I} = 2\frac{Z_{I+1}}{Z_I}\left(\frac{2\pi m_e kT}{h^2}\right)^{3/2} \exp\left(-\frac{\chi_I}{kT}\right)$$

where the Z_I are the partition functions for the corresponding atoms or ions and χ_I is the ionization energy to go from ion I to one fewer bound electrons.

8. Knowledge of the state of motion of light-emitting objects comes primarily from the Doppler shift

$$\frac{\mathbf{v}\cdot\hat{\mathbf{r}}}{c} = \frac{v_r}{c} = \frac{\lambda_{\text{obs}} - \lambda_{\text{rest}}}{\lambda_{\text{rest}}} = \frac{\Delta\lambda}{\lambda_{\text{rest}}}$$

The remainder of the book now takes the ideas we've developed on orbits and solving differential equations and brings them together with the ideas in this section on the interaction of light and matter, and explores some specific examples of astrophysical situations, particularly related to the surfaces and interiors of stars and planets.

Part III

Planetary and Stellar Structure

An Introduction to Astrophysics with Python
Stars and planets
James Aguirre

Chapter 19

Surface Temperatures of Planets

We put together ideas from our understanding of thermal radiation, a simple stellar model, and absorption and reflection to build a series of increasingly sophisticated mathematical models describing the surface temperatures of planets. We discuss the notion of model building and then apply it to the temperatures of planets in our solar system.

19.1 Overview

Having learned about blackbody radiation, we're now in a position to tackle an interesting question in astrophysics: how hot is the surface of planet? How does that depend on the planet's host star and the properties of the planet? These are obviously key questions for understanding whether a planet can hold an atmosphere[1] and whether it will be suitable for life.[2] And, we will find, we can make reasonably accurate predictions for planetary surface temperatures with a small amount of physics and some (reasonable) assumptions.

19.2 Background Assumptions

Let's start by making an oversimplified set of assumptions so that we can write down a physical model. A *physical model* is a set of equations describing some physical situation that allow us to calculate an interesting quantity or quantities. By making our assumptions explicit, we can then go back and see how to adjust that model quantitatively by adding in new features to the equations that define it.

---

[1] We explored this a bit in the exercise in Chapter 13.

[2] Which, as we currently understand it, requires liquid water on the surface, and thus a fairly narrow range of temperatures, the widest range being between 273 and 373 K.

doi:10.1088/2514-3433/ade5f6ch19

Here are the assumptions we will make about stars and planets to address the problem of determining the surface temperatures of planets:

1. Planets and stars are both perfect blackbodies at well-defined temperatures T_{planet} and T_{star}.
2. A planet absorbs radiation from the star that falls on the side of the planet facing the star, and re-radiates over its full surface.
3. The planet adjusts its temperature instantly, and the temperature is uniform across the planet.
4. Planets do not generate any internal heat that is important in determining their surface temperature.

Clearly, *every one of these assumptions fails to represent reality accurately*. You might not have even thought about Assumption 4, and actually verifying quantitatively that it is correct for any given planet could be quite difficult.[3] For now I will assert it, and we'll see how far that can take us. This is a general strategy taken by physicists: tackle problems by identifying the dominant effect first, allowing us get most of the answer for much less work, and then later adding in the more difficult effects that require more work to calculate.[4] We will see that will be able to make some reasonable adjustments to Assumptions 1, 2, and 3 and see how this changes our answer. Assumption 4 turns out to be quite good for the four inner, terrestrial planets (Mercury, Venus, Earth, and Mars), and not bad for the outer, gaseous planets (Jupiter, Saturn, Uranus, and Neptune).

Making these assumptions lets us get right to writing down some equations. The flux from a star of luminosity L at a distance r from the planet is just

$$\boxed{F = \frac{L}{4\pi r^2}} \tag{19.1}$$

A planet of radius R presents a (cross-sectional) area of

$$A_{abs} = \pi R^2 \tag{19.2}$$

(the area of the projected circle) over which this flux is absorbed and an area

$$A_{rad} = 4\pi R^2 \tag{19.3}$$

over which it radiates power out (Assumption 2). Thus, the power falling upon the planet is

$$P_{in} = A_{abs}\frac{L}{4\pi r^2} = \pi R^2 \frac{L}{4\pi r^2} \tag{19.4}$$

---

[3] It's not obvious, but it does turn out that star's radiation is almost always the dominant determinant of the surface temperature, but rarely determines the *internal* temperature, at least for planets in our solar system.
[4] Sometimes, a problem just can't be broken apart this way, such as when many equally important competing effects are present, or two large effects are nearly balanced, and it isn't clear which is dominant. Then we just have to tackle a hard problem.

and the power being radiated out is

$$P_{\text{out}} = A_{\text{rad}}\sigma T_{\text{planet}}^4 = 4\pi R^2 \sigma T_{\text{planet}}^4 \qquad (19.5)$$

Now, by assuming that in thermal equilibrium, $P_{\text{in}} = P_{\text{out}}$, we have our model:

$$T_{\text{planet}} = \left(\frac{1}{4\sigma}\frac{L}{4\pi r^2}\right)^{\frac{1}{4}} \qquad (19.6)$$

19.3 Refining the Model

Now, we know there are two interesting ways in which Assumption 1 is wrong: planets do reflect some light (we can see them, after all!), and the light which is reflected can't go into heating them up. Moreover, the (mostly visible) light from the star is at a different wavelength than the (mostly infrared) light radiated by the planet, and the planet need not be equally black at both wavelengths. For example, you may know of the greenhouse effect, which occurs because the atmosphere is more transparent to visible light than to infrared.

Since the incident flux can only be absorbed or reflected, we can write the total as the sum of these fluxes

$$F = F_{\text{ref}} + F_{\text{abs}} \qquad (19.7)$$

We define the fraction of light reflected as the *albedo*, which we'll denote α, and clearly this reduces the initial input flux. This definition is

$$\alpha = \frac{F_{\text{ref}}}{F} \qquad (19.8)$$

so that

$$\frac{F_{\text{abs}}}{F} + \frac{F_{\text{ref}}}{F} = 1 = \frac{F_{\text{abs}}}{F} + \alpha \qquad (19.9)$$

or

$$F_{\text{abs}} = (1 - \alpha)F = (1 - \alpha)\frac{L}{4\pi r^2} \qquad (19.10)$$

where $0 < \alpha < 1$, and $\alpha = 0$ for a perfect blackbody. This means the input power that can go into heating the planet is smaller than if it were a perfect blackbody:

$$P_{\text{in}} = (1 - \alpha)A_{\text{abs}}\frac{L}{4\pi r^2} = (1 - \alpha)\pi R^2 \frac{L}{4\pi r^2} \qquad (19.11)$$

Usually we think about the albedo as being a property of the material on the planet's surface (or in its atmosphere): a planet covered in snow will be more reflective than one covered in black dirt, for example. Effects that have to do with the fact that not all of the area πR^2 is being illuminated in the same way can be included by letting A_{abs} deviate from the geometrical value; more on this below.

Now, we're going to allow for the planet to be an imperfect radiator as well. The details of how the atmosphere (for example) fails to be a blackbody are complicated, but we can encapsulate the effect into a single number $0 \leqslant \beta \leqslant 1$, since any imperfect radiator will radiate *less* than a blackbody at the same temperature. The quantity β is referred to as the *emissivity*. So, for the planet, we will write

$$P_{\text{out}} = \beta A_{\text{rad}} \sigma T_{\text{planet}}^4 = \beta 4\pi R^2 \sigma T_{\text{planet}}^4 \tag{19.12}$$

Notice that while we've moved away from assuming perfect blackbodies, for an equilibrium situation we still must have energy conservation and $P_{\text{in}} = P_{\text{out}}$. The actual temperature will depend on both α and β, according to our new model

$$T_{\text{planet}} = \left(\frac{1-\alpha}{4\sigma\beta} \frac{L}{4\pi r^2} \right)^{\frac{1}{4}} \tag{19.13}$$

Notice that $\alpha > 0$ *decreases* the temperature, and $\beta < 1$ *increases* the temperature.

You'll notice that in order to make the model better, we've had to add some complication. We can measure α and β for the Earth or nearby planets, but they are hard to calculate without detailed information on planetary surfaces and atmospheres, so it's hard to get good predictions for what they might be for extrasolar planets.

Finally, let's look at Assumptions 2 and 3. We know that the planet is actually a sphere, so the absorption normal to its surface doesn't actually result in an absorbing area of πR^2. The fact that heat can be distributed (by conduction and convection), and the planet can rotate and has some heat capacity, means that the temperature isn't uniform. We might choose to consider a non-uniform T to incorporate these realities, but it will be easier to simply say that, for some average temperature T, the absorbing and emitting areas are modified from their geometrical value to some "effective" value to reflect the amount of radiation actually emitted, thus:

$$T_{\text{planet}} = \left(\frac{1}{\sigma} \frac{A_{\text{abs,eff}}}{A_{\text{rad,eff}}} \frac{1-\alpha}{\beta} \frac{L}{4\pi r^2} \right)^{\frac{1}{4}} \tag{19.14}$$

You will notice something interesting about our model: the planet temperature ends up depending on the product

$$\xi = \frac{A_{\text{abs,eff}}}{A_{\text{rad,eff}}} \frac{1-\alpha}{\beta} \tag{19.15}$$

Since I have four possible quantities to adjust, but just one number ξ that ends up affecting T_{planet}, it's clear that even at the same L and r it is possible to have different values of T_{planet}: it's not just how far away that matters. If I make β small enough, I can make the planet very hot, and if I make α large enough, I can make it very cold. I can also play with the speed of rotation and the heat capacity to change the ratio of effective emitting and absorbing areas. This also makes it possible to have planets with the same temperature, but very different distances, if they have very different planetary properties, as expressed in the *parameters* α, β, $A_{\text{abs,eff}}$ and $A_{\text{rad,eff}}$.

In fact, now you can see that my model is in some ways *too* flexible: I can make T_{planet} be pretty much anything by adjusting the four numbers above, and as noted, a wide range of parameters will lead to the same effect in Equation 19.15. We say the model is *degenerate* in these parameters, and we need additional information to come to a unique solution. This additional information can come from a direct measurement of some other property of the planet (its spectrum or rotation or composition) that allows us to determine directly what some or all of the parameters are, or at least bound the range of possibilities.

In particular, we have lumped all of the interesting physics of the greenhouse effect into the parameter β, which does not make it clear how the greenhouse effect actually raises the surface temperature. The key idea is that the atmosphere is (mostly) transparent to visible light from the Sun, but much more opaque to infrared light being radiated from the surface. A more realistic model for the greenhouse effect would break the "surface" into several layers, including the actual surface at the bottom and several layers of atmosphere. The net effect is that the upper parts of the atmosphere appear to radiate back to space at a lower temperature (hence $\beta < 1$) than the actual surface, where the additional heat has been trapped.

19.4 Exercise: Surface Temperatures in the Solar System

The goals for this exercise are
- Appreciate the role of successive approximation in solving a scientific problem: what basic physics gets most of the answer correct? How do we improve the correctness by adding additional complexity or realism?
- Use our knowledge of blackbody radiation to construct a simple model of planet surface temperatures
- Quantitatively compare our model to actual data
- Explore what additional features of the model can be made more realistic and improve the agreement between theory and observation

Table 19.1 lists data for the 8 planets of our solar system (taken from Wikipedia, but checked against NASA), as well as the dwarf planets Ceres and Pluto. Recall

Table 19.1 Parameters of the Solar System Bodies Relevant for Surface Temperature Calculations

Planet Name	Distance from Sun (AU)	Surface Temperature (K)	Bond Albedo (fractional)
Mercury	0.387	440	0.068
Venus	0.723	737	0.900
Earth	1.000	288	0.306
Mars	1.524	210	0.250
Ceres	2.770	168	0.034
Jupiter	5.203	165	0.343
Saturn	9.539	134	0.342
Uranus	19.180	76	0.300
Neptune	30.060	72	0.290
Pluto	39.480	32	0.720

that the inner four planets are "rocky", that is, they have primarily metal (iron) and rock (silicon, oxygen) compositions, and the outer four are "gas giants", with the bulk of their mass coming from hydrogen and helium.

Question 1. Find a way to import the data in Table 19.1, being sure to add `astropy` units to the quantities as appropriate.

In the plots in the following questions, it will be more useful to give each planet a separate marker, rather than connecting the data points for each planet. (More plot symbols can be found here.) You can do this using the syntax below in the example.

It will also be helpful to label the planets with their names using `plt.text`. An example is shown in Figure 19.1.

Use a log–log plot (`plt.xscale'log'`)), and adjust the axes to show all the planets clearly (and also to allow you to read off the axis labels). What is the general trend with distance? Do any planets go against the general trend?

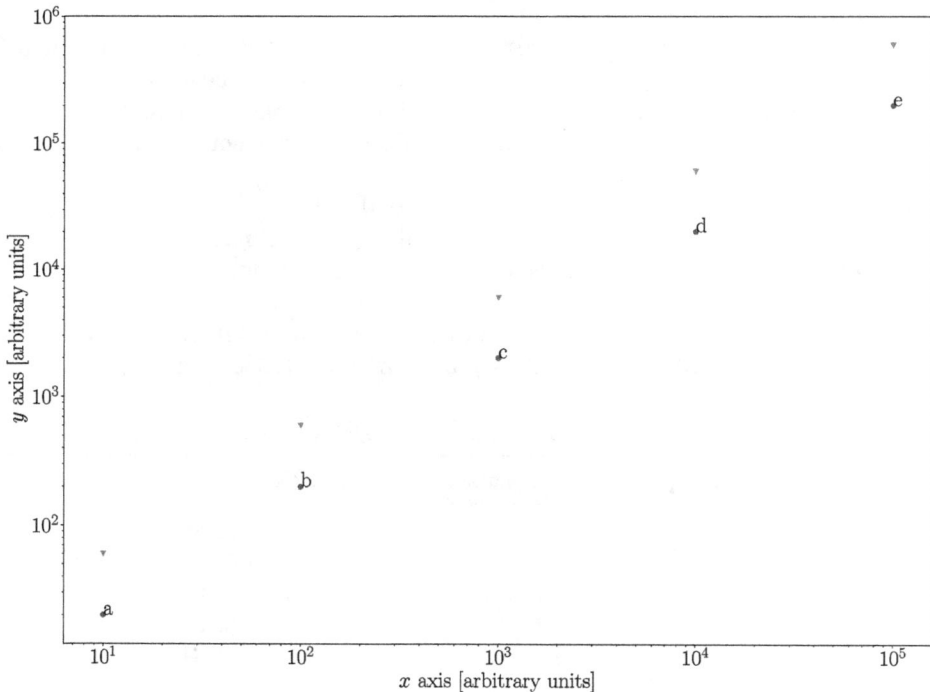

Figure 19.1. An example of the use of `plt.text`.

```
# Labeled plot example
x = np.logspace(1, 5, num=5)
y = 2*x
names  = ['a', 'b', 'c', 'd', 'e']
plt.plot(x, y, color = 'blue', marker = 'o', linestyle = '
   None') # plot blue circles


plt.plot(x, 3*y, 'rv') # plot red downward pointing
   triangles; note shortcut notation
plt.xscale('log')
plt.yscale('log')
for i, name in enumerate(names):
    plt.text(x[i], y[i], name)
```

Question 2. The full model for the surface temperature of a planet is given by Equation 19.14. Write a Python function that implements this model by finding the surface temperature given a stellar luminosity L, distance from star to planet d, albedo α, greenhouse factor β, and calculates the planet's temperature.

Let's call Model 1 the simplest case, which has $\alpha = 0$, $\beta = 1$, $A_{\mathrm{abs,eff}} = \pi$ and $A_{\mathrm{rad,eff}} = 4\pi$.

Make a new plot, which has the true values, as well as Model 1. In this case, plot Model 1 as a continuous line so you can more easily see the trend (the only thing specific to the planet that this model depends on is the distance d).

Qualitatively, do the surface temperatures follow the trend? Calculate the actual prediction of Model 1 for all the planets in our solar system, and compare to the true values in the table. What is the relative error $T_{\mathrm{model1}} / T_{\mathrm{true}}$?

Question 3. Let's call the next most sophisticated model, including the albedo, Model 2.
- Make a new plot with the true values, the line for Model 1, and the Model 2 values, plotted as markers with a different color or symbol.
- Calculate again the relative errors. For which of the planets does Model 2 do a better job of predicting their temperatures? Does Model 2 make *larger* errors than Model 1 for any planets?

Question 4. Since we still have differences between the true values and Model 2, there must be something besides the albedos that are important, so we need to seek another way to make our model agree better. We will change the efficiency of re-radiation, the so-called *emissivity* β. This leads to Model 3.

For two of the inner, rocky planets, you should find that Model 2 gives a poor estimate of their temperature. Which two planets are these, and what values of β are necessary get Model 3 to give the correct surface temperatures? Do you know anything about these planets that makes sense of why they require adjustment to their blackbody radiation via β?

Question 5. For the outer four gas giant planets, suppose I tell you that there is no physical reason to adjust the β, A_{rad} or A_{abs} parameters in the model. What assumption of our model is likely being violated? What are some possible additional explanations for the discrepancy between the model and the true temperatures?

19.5 Study Questions

1. Let's consider some outside, alien observers of our solar system, and see how they might be able to observe the Earth and Jupiter in the light of our Sun. Their viewing situation is shown in Figure 19.2. In the following, assume the

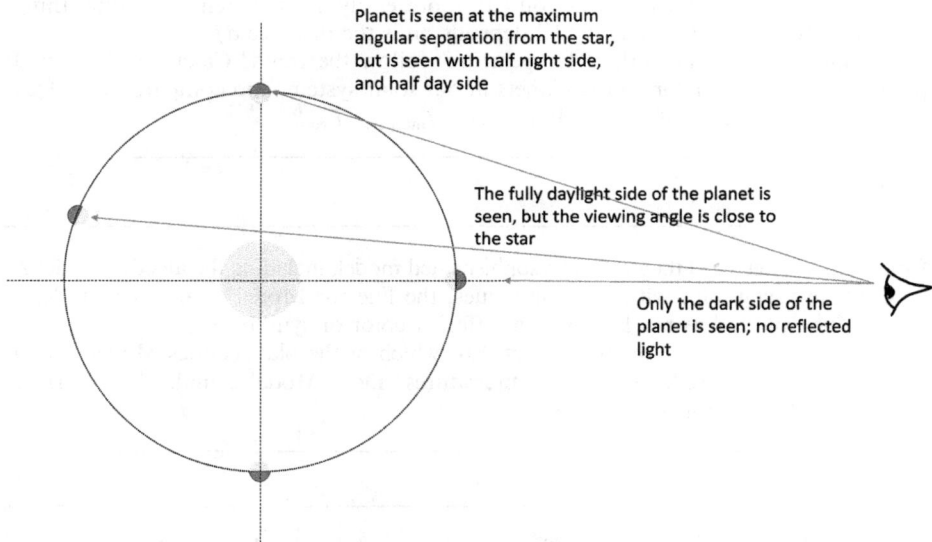

Figure 19.2. An illustration of why an observer outside the solar system would most likely to be able to see reflected sunlight from only half of a planet's surface. In order to see the fully illuminated face of the planet, the planet needs to be nearly "behind" the star from the point of view of the observer, and therefore very hard to distinguish in the glare of the star. When it appears at the largest angular separation, it would be seen it as about half illuminated.

characteristics of these objects given below (note that the temperature and albedo of the planets are the same as we assumed in Table 19.1).

Object	Surface Temperature (K)	Albedo	Distance from Sun (AU)	Radius (m)
Sun	5800	—	0	6.96×10^8
Earth	288	0.306	1	6.38×10^6
Jupiter	165	0.343	5.203	7.15×10^7

(a) Assume the Sun, Earth, and Jupiter are perfect thermal emitters, and that all of them radiate at the given surface temperature.
  i. What is the average energy of a photon emitted by each of these objects?
  ii. What are the wavelengths of light corresponding to these energies?
  iii. What part of the electromagnetic spectrum are these wavelengths in?

(b) The Earth and Jupiter will reflect sunlight back into space, with an average photon energy equal to the Sun's average photon energy (they're basically just acting as mirrors, which don't change the wavelength). The higher the albedo, the greater the reflected light.
  i. Calculate the flux (Wm^{-2}) of sunlight *incident* on the Earth and on Jupiter. (This is two different values.)
  ii. Calculate the *luminosity*, in Watts, in reflected light, of the Earth and Jupiter. Assume the relevant surface area for calculating the luminosity reflected back into space is *half* the projected area of the planet (i.e., its circular area); see Figure 19.2.

(c) What is the ratio between the bolometric luminosity of *reflected* light for the Earth and Jupiter and the luminosity of the Sun?

(d) ⋆ One of the problems of detecting extrasolar planets is seeing them in the light of their star. Let's compare the thermal and reflected *spectra* to see if there might be specific wavelengths where the planets are relatively brighter compared to the light from the star. To fairly compare, we'll put the spectra all in units of $W\ m^{-1}$.
  Put on a single plot (log–log axes):

  • The thermal spectra of the Sun, Earth, and Jupiter. These all have the form $A_{\text{object}}\,\pi B_\lambda(T_{\text{object}})$ where A_{object} is the surface area of the object.
  • The reflected spectra of the Earth and Jupiter. These will all have the spectral "shape" given by $B_\lambda(T_{\text{sun}})$, since it is the Sun's light

that is reflected. However, they will also need to account for the planet's albedo and distance from the Sun; thus their spectra look like (albedo and distance factors) $A_{\text{reflective}} \times \pi B_\lambda(T_{\text{sun}})$ where $A_{\text{reflective}}$ is the area that is reflecting light (see above).

If the goal is to minimize the ratio between the brightness of the Sun and the planet, for what range of wavelengths does this occur?

An Introduction to Astrophysics with Python

Stars and planets

James Aguirre

Chapter 20

Hydrostatic Equilibrium

After a long hiatus, we return to the subject of gravitation and how it affects large objects, both with respect to their internal structure and their atmospheres. We develop two differential equations to describe the distribution of mass and the variation of pressure with position, which will turn out to be critical in describing stellar structure in Chapter 23.

In this section, we're going to begin our consideration of the "structure" of the atmospheres and interiors of stars and planets. By this we mean the way in which temperature, density, and pressure of the matter varies as a function of distance from the center of the star or planet, or as a function of height in the atmosphere. The key issue is that we are considering objects that are massive enough that the force of gravity has a significant impact on their structure.[1] The question we want to consider is what forces balance the inward pull of gravity so that we reach an equilibrium situation with the atmosphere or the object neither expanding or contracting. We'll start with a simple one-dimensional (1D) model of an atmosphere and then discuss generalizing that idea to a spherical object, like a star or planet. But even in the spherical case, we're only going to consider the interesting structure to occur in the radial direction (another 1D model).

20.1 Stellar and Planetary Atmospheres (Slab Geometry)

We know that gravity holds Earth's atmosphere to it, and for this case there's a well-defined lower boundary (the Earth's surface) and a less well-defined upper boundary (where the atmosphere "ends" and space begins). Similar considerations would

---

[1] Unlike, say, you, whose structure is completely determined by electromagnetic forces.

doi:10.1088/2514-3433/ade5f6ch20   20-1

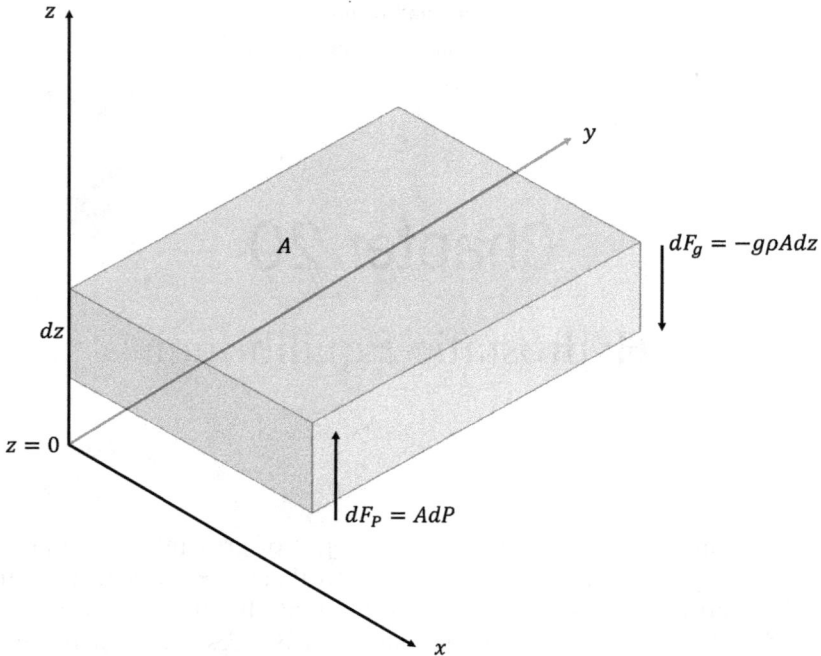

Figure 20.1. The geometry of a slab of material in hydrostatic equilibrium, with gravitational forces acting in the $-z$-direction and a pressure gradient resisting gravitational pull in the $+z$-direction.

apply to Venus, or any terrestrial planet large enough to hold an atmosphere.[2] For the outer gaseous planets and for stars, the notion of what you would call the "atmosphere" is a little more murky, since there is no solid surface. In practice, we define the atmosphere in this case to be the portion of the body where the gas or plasma stays relatively transparent; the atmosphere ends when you can no longer "see into" the body. We will take a look at how to make that idea more quantitative in later chapters (Figure 20.1).

We'd now like to look a little more carefully at the way in which gravity shapes the pressure and density of the atmosphere. Explicitly, we're going to try to find a relation $P(z)$ for the pressure at an altitude z in an atmosphere. We lay out a coordinate system in which $z = 0$ at the "surface" (either the actual surface, or the place where the atmosphere becomes opaque), and increases as we go up. We can then write down the relation for adding a little slab of mass dm as we move a small amount dz through the atmosphere:

$$dm = \rho dV = \rho(z)A dz \qquad (20.1)$$

---

[2] As we learned in the exercise on the surface temperatures of planets, Mercury, Ceres, and Pluto have no appreciable atmosphere; it turns out neither does the Moon. Mars does have some atmosphere, but it has relatively little effect on the surface temperature, so it wasn't apparent in that exercise.

where ρ is the mass density of the gas (which can be a function of height), and A is the area of a column of this gas, so that Adz represents a small change in the volume dV.

If we now consider the small change in force on this mass exerted by the gravitational field of the star or planet as we make a small change in height, we can write

$$dF_g = -gdm = -g\rho(z)Adz \tag{20.2}$$

where g is the gravitational acceleration acting on the little mass dm, and we assume the force is acting "inward" on the atmosphere, in the opposite direction that z increases. If we assume that the atmosphere at $z = 0$ starts a distance R from the center of the body of mass M, then

$$g \approx G\frac{M}{R^2} \tag{20.3}$$

The approximation comes from assuming that the atmosphere adds a negligible amount to M, and that the thickness of the atmosphere $h \ll R$. For the Earth's atmosphere, we can use the familiar $g = 9.8$ ms^{-2}, but for other objects we need to compute the *surface gravity g* appropriate to their mass M and radius R.

To oppose the force of gravity, we need another force, and we're going to look to the *pressure* in the material to provide it. For the moment, we're not going to consider *how* the pressure in the material is generated: it could be the internal pressure of a gas, or the pressure of radiation, or the pressure from the quantum nature of matter. (We will discuss all of these in turn later on.) We are simply going to suppose that some form of pressure is going to provide the counteracting force to gravity. Since a *pressure P* is just a force per unit area, we can conclude the change in pressure necessary to balance the gravitational force equal and opposite to the gravitational force

$$\frac{dF_P}{A} = dP = -g\rho(z)dz \tag{20.4}$$

and finally, we get a differential equation for the pressure as a function of height z:

$$\frac{dP}{dz} = -g\rho(z) \tag{20.5}$$

Note that this just rearranges the differential elements of Equation (20.4). We need to specify the initial condition

$$P(z = 0) = P_0 \tag{20.6}$$

which is the pressure at the "bottom" of the atmosphere. (For example, in the case of Earth's atmosphere, we would set P_0 to be the pressure at sea level.)

Now this is all well and good, but even if we're willing to assume that g is constant, we don't know $\rho(z)$, so it's not obvious how to integrate this differential equation. If we can treat the atmosphere as an ideal gas, then we can simplify things further by using the ideal gas law. We'll write it in the form

$$P = \frac{N}{V}kT = nkT = \frac{\rho}{\bar{m}}kT \tag{20.7}$$

where N is the total number of particles in a volume V, n is the number density of particles, ρ is the mass density, and $\bar{m}$ is the mean mass per particle in the gas.[3] Then we can write

$$\rho(z) = \frac{\bar{m}(z)}{k}\frac{P(z)}{T(z)} \tag{20.8}$$

since of course the temperature can depend on the altitude as well, and indeed the mean molecular mass could change with altitude if the composition of the atmosphere is changing with height. At first glance, this doesn't seem to have made things better, since we don't know $T(z)$ either. But it turns out that assuming T is constant is a much better assumption, for example, that assuming that ρ is constant or known. (In the case of the Earth's atmosphere, from the surface to an altitude of 120 km, the temperature varies between about 180 and 300 K, less than a factor of 2 variation. By contrast, the density varies by 7 orders of magnitude over that range. It turns out that the temperature evolves rather slowly with height in the Sun's atmosphere as well.) For now, we'll also assume that the mean molecular mass is constant with height as well.

Making these assumptions, we have

$$\frac{dP}{dz} = -\frac{\bar{m}g}{kT}P(z) \tag{20.9}$$

By dimensional analysis, the quantity

$$\Lambda \equiv \frac{kT}{\bar{m}g} \tag{20.10}$$

has units of length, and we term it the *scale height*. By our assumptions, it is also a *constant*. The physical interpretation becomes clear if we solve the differential equation

$$\frac{dP}{dz} = -\frac{P}{\Lambda} \tag{20.11}$$

Using our initial condition Equation (20.6), the solution is just

$$\int_{P_0}^{P} \frac{dP}{P} = -\int_{0}^{z} \frac{dz}{\Lambda} \tag{20.12}$$

---

[3] This can be a bit of a confusing concept for gases where there are a mix of different kinds of particles; we'll take this up more carefully at the end of the chapter in Section 20.3.

$$\ln\left(\frac{P}{P_0}\right) = -\frac{z}{\Lambda} \tag{20.13}$$

or

$$P(z) = P_0 \exp(-z/\Lambda) \tag{20.14}$$

So the scale height is the distance over which the pressure drops by a factor of $1/e$.

20.2 Stellar and Planetary Interiors (Spherical Geometry)

Equation (20.11) is a useful approximation for the Earth (or Sun) because their atmospheres are thin and negligible in mass compared to the mass of the bulk of the Earth (or Sun), and because g, T and $\bar{m}$ are pretty constant. However, if we want to talk about pressure balance for an entire spherical object, like the bulk of the Sun, we need to go back and look at Equation (20.2) again.

That derivation may have obscured the role that force balance plays in the process, so let's go back a bit more carefully. Dropping the assumption that g is independent of distance, we can write the small change in gravitational force on a small change in mass at a distance r from the center of a spherically symmetric body as

$$dF_g = -G\frac{M_r}{r^2}dm \tag{20.15}$$

where we have defined the mass enclosed inside a radius r as

$$M_r = \int_0^r \rho(r')4\pi r'^2 dr' \tag{20.16}$$

The 4π is present because we have assumed complete spherical symmetric in the other coordinates θ and ϕ, so I need only consider the surface area element $4\pi r^2$ times the radial change dr to get a volume. It is an interesting and important theorem that the gravitational force of any spherically symmetric mass *outside* the radius r does not contribute to the net gravitational force on dm. Again using

$$dm = \rho(r)Adr \tag{20.17}$$

we get

$$dF_g = -G\frac{M_r}{r^2}\rho(r)Adr \tag{20.18}$$

If this gravitational force were unopposed by any other forces, then the spherical mass would continue to contract, and we would not have a static solution where the radius is not changing with time, and clearly the radius of the Sun and Earth are pretty constant. So something must be opposing the inward crush of gravity. Equation (20.11) suggests the solution: we need a pressure *gradient* to balance the gravitational force. We can write the radial component of pressure gradient (assuming, again, no dependence on θ and ϕ) as

$$\frac{dP}{dr} \equiv \frac{1}{A}\frac{dF_P}{dr} \tag{20.19}$$

or

$$dF_P = AdP \tag{20.20}$$

To achieve equilibrium, we need the *total* force on the mass element to be zero, with gravity acting in one direction and the pressure gradient in the other:

$$dF = 0 \tag{20.21}$$

$$= dF_g - dF_P \tag{20.22}$$

$$= -G\frac{M_r}{r^2}\rho(r)Adr - AdP \tag{20.23}$$

Canceling the common area factor on each side and rearranging, we can balance the gravitational force inward if we have a pressure gradient (generated by means yet to be specified), which satisfies

$$\frac{dP}{dr} = -G\frac{M_r}{r^2}\rho(r) \tag{20.24}$$

Equation (20.24) expresses the key idea behind stability against gravitational collapse: not just a pressure, but a pressure *gradient* must oppose the inward pull of gravity. How this pressure gradient is generated is not specified, and it will be different for stars and planets. In the case of stars, it is one of the so-called *stellar structure equations*. It is conventional that we define a second structure equation from Equation (20.16), by differentiating that equation:

$$\frac{dM_r}{dr} = 4\pi r^2 \rho(r) \tag{20.25}$$

If we define the total mass of the object as $\mathcal{M}$ and its radius as $\mathcal{R}$, then clearly

$$\int_0^{\mathcal{R}} M_r dr = \int_0^{\mathcal{R}} 4\pi r^2 \rho(r)dr = \mathcal{M} \tag{20.26}$$

Now, if we were magically given $\rho(r)$, we could solve Equation (20.25) and then (20.24) relatively easily. Equation (20.26) implies the initial condition for $M_r(r)$ is $M_r(0) = 0$. If we also know $P(0) = P_0$ and $P(\mathcal{R}) = P_s$, a complete solution can be specified. Unfortunately, even in this case we rarely know exactly what P_0 and P_s should be, though $P_s \approx 0$ is required so that the "surface" of the object is static (no pressure is pushing it outward). For the materials of which planets are made, there is usually some relation between P and ρ, called an *equation of state*, which helps simplify this problem, though setting the initial conditions is still tricky.[4]

---

[4] A relation between ρ and P can also be established for more general cases, leading the theory of so-called "polytropes", which can be useful approximations to stellar interiors.

In the case of stars, however, it turns out to be harder yet to solve these equations, since we will find that $\rho(r)$ also depends on the flow of heat (from nuclear fusion) through the star, so in addition to Equation (20.24) for the pressure and Equation (20.25) for the enclosed mass, we are also going to need differential equations for the change in *temperature* with radius, as well the flow of energy released by fusion.

20.3 Mean Molecular Mass

Let's go back and revisit the *mean mass per particle* $\bar{m}$ that we introduced back in Equation (20.8). The thing we are trying to do is describe, in a single number, the average mass of all the different kinds of particles we might have in a gas. Suppose we have N total particles, which could be of all different types, with a total mass M in a volume V. Then conceptually the mean mass per particle in that volume is very simple

$$\bar{m} = \frac{M}{N} \tag{20.27}$$

Since the mean density is

$$\rho = \frac{M}{V} \tag{20.28}$$

and the mean number density is

$$n = \frac{N}{V} \tag{20.29}$$

we could also write

$$\bar{m} = \frac{\rho}{n} \tag{20.30}$$

More commonly, however, we are not given M and N (or ρ and n), but rather the relative fractions of different types of particle, either by *number* or by *mass*, and the mass per particle for each type. How would we go about computing $\bar{m}$ in that case?

First, let's enumerate each type of particle (atoms, molecules, ions, electrons) with an index j, and note that the total number N is just the sum over all the types:

$$N = \sum_j N_j \tag{20.31}$$

where N_j is number of type j. Dividing through by the volume they occupy gives

$$n = \sum_j n_j \tag{20.32}$$

Dividing through by total number density leads to

$$1 = \sum_j \frac{n_j}{n} = \sum_j \frac{N_j}{N} \tag{20.33}$$

and we call the quantity

$$f_j = \frac{n_j}{n} = \frac{N_j}{N} \qquad (20.34)$$

the *fraction by number*, or sometimes just the *fraction*, of particles of type j. This just expresses that if you add up the fraction of the total for each particle type, you must get 1.

We can write similar equations for the mass, namely

$$M = \sum_j M_j = \sum_j m_j N_j \qquad (20.35)$$

where M_j is the total mass of particles of type j, and m_j is the mass per particle for that type. Then we can easily see that

$$\rho = \sum_j \rho_j = \sum_j m_j n_j \qquad (20.36)$$

and

$$1 = \sum_j \frac{M_j}{M} = \sum_j x_j \qquad (20.37)$$

where x_j is the *mass fraction* of type j.

Our goal is now to get expressions for $\bar{m}$ in terms of the m_j (the individual masses per particle) and either the f_j (the fraction by number) or x_j (the fraction by mass). The fraction by number is easier, since we can easily write down

$$\bar{m} = \frac{M}{N} = \frac{\rho}{n} = \frac{1}{n}\sum_j m_j n_j = \sum_j m_j \frac{n_j}{n} = \sum_j m_j f_j \qquad (20.38)$$

The mass fraction requires a little more cleverness. Since we need to write in terms of the mass fraction

$$x_j = \frac{M_j}{M} \qquad (20.39)$$

we could notice that

$$\frac{1}{\bar{m}} = \frac{N}{M} = \frac{1}{M}\sum_j N_j = \sum_j \frac{N_j}{M} = \sum_j \frac{N_j}{M_j}\frac{M_j}{M} \qquad (20.40)$$

where the last step just multiplies each term in the sum by M_j / M_j. But now we have

$$\frac{1}{\bar{m}} = \sum_j \frac{N_j}{M_j}x_j = \sum_j \frac{N_j}{m_j N_j}x_j = \sum_j \frac{1}{m_j}x_j \qquad (20.41)$$

It is common to define a dimensionless *mean molecular weight*,[5] which measures $\bar{m}$ in units of the hydrogen atom m_H as

$$\mu = \frac{\bar{m}}{m_H} \tag{20.42}$$

so that we have

$$\mu = \sum_j \frac{m_j}{m_H} f_j \tag{20.43}$$

and

$$\frac{1}{\mu} = \sum_j \frac{m_H}{m_j} x_j \tag{20.44}$$

20.4 Exercise: Atmospheres and Interiors

The goals of this exercise are to
- Understand how gravity influences the "structure" of an atmosphere. By "structure" here we mean the dependence of density and pressure on height above the planet's surface. (It could also mean how the temperature varies, but in this exercise we are going to assume an isothermal atmosphere.)
- Understand the generalization of the idea of *hydrostatic equilibrium* to spherical, self-gravitating objects
- Understand the first two *stellar structure equations* describing the internal structure of a star

---

Question 1. Let's calculate the dimensionless mean molecular weights μ of the following situations. (Parentheses following an element give the atomic mass number to use).
- the atmosphere of the Earth: 78% N_2 (28), 21% O_2 (32), 1% Argon (40) *by number*
- the atmosphere of a star with only H (1) and He (4). Of the atoms, 90% are H and 10% He, *by number*, but 50% of the hydrogen is ionized, and none of the He is ionized.

---

[5] Which gets a failing grade as a name, since (a) it need not refer to molecules, and (b) it is not a weight. It does get some points for being a mean.

Question 2. Using the mean molecular mass for the Earth's atmosphere you calculated in Question 1, and assuming a typical temperature of 300 K, what is the scale height of the Earth's atmosphere? (Write a function: you're going to calculate more scale heights.)

Assuming our simple model for the pressure with altitude, how much lower is the pressure than at sea level for

- Mt. Everest (latest elevation 8848.86 m)
- the air outside a commercial airliner (30,000 ft altitude)
- a satellite in low earth orbit 200 km above the surface

Question 3. The stellar atmosphere has a temperature of 5800 K and a mean molecular weight of 0.62.

Calculate the gravitational acceleration at the surface of the Sun.

How does the product $\bar{m}g$ for the Sun compare to that for the Earth? (Note: the g's are not the same!)

What is the scale height of the Sun's atmosphere? How does it compare to Earth's?

Question 4. The material of which the Earth is made (rock, metal) has a nearly constant density, independent of the pressure: we say that it is *incompressible* (or close to it). Thus, though the pressure increases toward the center of the Earth, the density does *not*.

Suppose we make a model of the interior of the Earth, assuming the density to be constant (independent of the distance from the center of the Earth), denoted ρ_0. Find the mass enclosed $M_r(r)$ in a radius r for $0 \leqslant r \leqslant \mathcal{R}$, where $\mathcal{R}$ is the radius of the Earth. Your expression will depend on ρ_0. Re-write ρ_0 in terms of the total mass $\mathcal{M}$ and $\mathcal{R}$ and re-express M_r in terms of these variables.

Question 5. Again assuming a total mass $\mathcal{M}$ and radius $\mathcal{R}$ for a constant density ρ_0, find the pressure $P(r)$ necessary to support this. Note that $P(r)$ will *not* be constant, even though we are assuming the rock does not change density. You should be able to write the equation in such a way that you find $P_c - P(r)$ is equal to an expression you can integrate. Here, $P_c = P(r = 0)$, or the central pressure. To find an expression for P_c, assume that $P(\mathcal{R}) = 0$.

Question 6. Make plots of $M_r(r)$ and $P(r)$ for the Earth (putting in appropriate values for $\mathcal{M}$ and $\mathcal{R}$). Try plotting these both with linear axes, and log–log.

Question 7. Use the expression for the central pressure from Question 3 to estimate the central pressure of the *Sun*, if it were a sphere of constant density equal to its average density. For the Sun, if we assume the ideal gas law, we can also estimate the central *temperature*, assuming the average density and a mean molecular weight of 0.62. Note that in this case, we're going to severely underestimate the central pressure, because the material of which the Sun is made is not incompressible.

20.5 Study Questions

1 Consider a star of mass $M = 2.7 M_{Sun}$ and radius $R = 2.31 R_{Sun}$. The atmosphere of this star has a surface temperature of $T = 10, 100$ K and has a mean molecular mass of $\mu = 1$. The cross-section for the interaction of a photon (of any wavelength) with a free electron is $\sigma_T = 6.65 \times 10^{-29}$ m^2. The cross-section for interaction with a hydrogen nucleus (proton) is much smaller and can be neglected in the following. We will consider wavelengths well away from the hydrogen *atom* energy levels, so the total cross-section of the atmosphere to a photon is just σ_T.

  (a) What is the scale height of the atmosphere of this star?

  (b) If the number density of electrons in the atmosphere is 8×10^{21} m^{-3}, what is the optical depth of one scale height of this atmosphere? (Assume the temperature and number density of electrons do not change over one scale height; we'll see what happens when the number density does below.)

  (c) Under the assumptions above, by what fraction does light passing through this atmosphere of free electrons get attenuated?

  (d) Let's now assume the number density of electrons scales with the density in the same way as the density of an atmosphere with height: $n_e(x) = n_0 \exp(-z/H)$ where H is the scale height and $n_0 = 8 \times 10^{21}$ m^{-3}. What is the attenuation of the light passing through, assuming this profile for the free electrons? To do this, we should go back to Equation (17.13) to write the differential equation for the change in flux while going through a slab as

$$\frac{dF}{dx} = -\sigma_T n_0 \exp(-x/H)F \qquad (20.45)$$

to get the partially integrated form

$$\ln\left(\frac{F}{F_0}\right) = -\int_0^\infty \sigma_T n_0 \exp(-x/H)dx \tag{20.46}$$

How much does your answer change from part (c) by considering this more realistic density profile?

An Introduction to Astrophysics with Python
Stars and planets
James Aguirre

Chapter 21

The Main Sequence and the Hertzsprung–Russell Diagram

We introduce the "main sequence" and the importance of the stellar mass in determining of the properties of stars. We briefly discuss some ideas in star formation and the role of nuclear fusion and the implications for the lifetimes of stars. We introduce simple relations between stellar mass and the main sequence luminosity, radius, and temperature to explore some features of the main sequence.

21.1 Key Ideas

The goal of this chapter is to make clear the importance of the mass M of a star (when it is initially formed) in defining (nearly) all of its properties. We are going to consider the empirical evidence that there is a nearly one-to-one relation between a star's mass at its formation and its observable quantities, particularly the surface temperature T_*, bolometric luminosity L_*, and radius R_*. The relation between T_* and L_* is known as the Hertzsprung–Russell (H–R) diagram and is shown for stars whose masses have been determined by methods like those in Chapter 16[1] and in Figure 21.2. Note that there is a clear clustering of stars along a line from lower right to upper left: these stars are the main sequence, whose defining feature is that they are powered by the fusion of hydrogen into helium: four hydrogen nuclei, or protons, combine to form one helium nucleus of two protons and two neutrons. There will thus be a relation between a star's mass and its "lifetime"; that is, the amount of time the star can use its nuclear fuel to create its observed luminosity. This luminosity due to nuclear reactions also supports the star against gravitational collapse.

---

[1] That is, by finding stars in binary systems and using their observed orbital parameters and the two-body equation to work out their masses.

doi:10.1088/2514-3433/ade5f6ch21   21-1

21.2 Star Formation

We can understand something of why stars are far apart and why their mass is determined at their formation from the following simple argument. The densest regions of the "clouds" from which stars form will have number densities of about $n_{\text{cloud}} = 1 \times 10^3 \, \text{cm}^{-3}$. A stunning example of a star-forming region is shown in Figure 21.1.

Question 1. If the cloud is 70% hydrogen by mass, 28% helium by mass, and 2% all other elements having a mean molecular mass of 15.5 m_{p}, what is the mean molecular mass of the cloud?

Figure 21.1. A view of the " Pillars of Creation" star-forming region from the James Webb Space Telescope (JWST), with the color scale indicating intensity of various wavelengths of light in the near-infrared. Young stars are evident as orange and reddish objects inside the dusty "pillars" extending from lower left to upper right. Credits: Science: NASA, ESA, CSA, STScI, Image Processing: Joseph DePasquale (STScI), Alyssa Pagan (STScI), Anton M. Koekemoer (STScI)

> **Question 2.** Convert number density $n_{H,cloud}$ into a mass density $\rho_{H,cloud}$, and compute the radius of the sphere that would be necessary to contain the mass of the Sun. Compare this to the average mass density of the Sun
>
> $$\bar{\rho}_\odot = \left(\frac{4\pi}{3} R_\odot^3\right)^{-1} M_\odot \qquad (21.1)$$
>
> How many times denser is the Sun (on average) than the cloud from which it formed?

So generally speaking, a star uses up the material in its immediate neighborhood in forming, and as the surrounding material is *even less dense* than $n_{H,cloud} = 1 \times 10^4 \, cm^{-3}$, the space between stars is largely empty, and the star (with some caveats) doesn't gain mass over its lifetime. It is, of course, possible to lose it in various ways, but most of those ways in fact occur near the end of its life. So we can treat the starting mass as nearly fixed.

21.3 Stellar Lifetimes

Because a star's "fuel" for creating luminosity (from nuclear fusion) is its mass, its "lifetime" should be roughly

$$\tau \propto \frac{M}{L} \qquad (21.2)$$

In this context, "lifetime" means the amount of time it can have ongoing nuclear fusion. After it has used all the nuclear fuel available to it, something interesting will have to happen.[2] However, let's look more at Equation (21.2), which is only a proportionality and not even in the right units yet. Einstein tells us the energy associated with a mass m is

$$E = mc^2 \qquad (21.3)$$

where c is the speed of light. So now we have

$$\tau = K \frac{Mc^2}{L} \qquad (21.4)$$

which does have units of time if K is dimensionless.

> **Question 3.** Suppose the Sun could convert *all* of its mass into energy, and its luminosity stayed constant at the current value $L = 3.85 \times 10^{26} \, W$. How long could the Sun "live"?

---

[2] We won't cover it in this book, but the end states of stars include white dwarfs, neutron stars, and black holes.

In reality, the number in Question 1 is *way* off, and the Sun's lifetime is estimated to be about 10×10^9 yr (10 billion years … and we're already 4.5 billion through that!). Not all of the Sun's mass actually gets to participate in nuclear fusion, and in fact it is only the material deep in the Sun's core that takes part, which amounts to about 10% of the mass. Let's write Equation (21.4) as

$$\tau = f\epsilon \frac{Mc^2}{L} \tag{21.5}$$

with $f = 0.1$ (the fraction of the mass that is involved in fusion) and ϵ, the Sun's efficiency in converting mass into energy.

Question 4. Using the above facts, what is ϵ?

The actual nuclear fusion process that occurs in the Sun takes 4 protons and, through a rather complicated series of steps, turns them into a helium nucleus (two protons and two neutrons). Clearly a lot is going on there, but let's just look at the starting and ending points. The mass energy we start with is $4m_p c^2$ and, and the mass energy we end with is $m_{\text{He}} c^2$, the mass of the helium nucleus.

Question 5. For this problem we need to keep many significant figures; seven should do. The mass of a proton is $1.6726218 \times 10^{-27}$ kg and the mass of a helium nucleus is $6.6464998 \times 10^{-27}$ kg. Is the mass of 4 protons equal to a helium nucleus? How much do they differ? Express the fractional difference as

$$\eta = \frac{4m_p - m_{\text{He}}}{4m_p} \equiv \frac{\Delta m}{4m_p} \tag{21.6}$$

Question 6. Compare the value you got for η in Question 4 to the value of ϵ in Question 2. What does this suggest about where the "missing" energy in Question 5 goes?

21.4 Fitting Formulas and Scaling Relations

Very often in (astro)physics, we find that there is clearly a relation between two quantities, but there is not a good enough theoretical understanding of this relation to provide a physically motivated equation describing the relationship. Or, alternatively, the relation is well understood, but the detailed calculation of the relation is complicated, while the resulting functional dependence is fairly simple. In both these cases, it is desirable to be able to compress the complicated empirical relation into something more computationally tractable, and so we often seek some functional form that describes the relation. This functional form depends on both what we think the underlying independent variables are, as well as some parameters that control the shape of the function; in general, we don't demand that these parameters actually correspond to physically meaningful quantities, only that they allow the function to fit the data well.

In the following, we're going to use such simple "fitting functions" for the main sequence luminosity L and radius R as a function of the stellar mass M. These are given by Demircan & Kahraman (1991):

$$\frac{L(M)}{L_\odot} = \begin{cases} 0.35 \left(\dfrac{M}{M_\odot} \right)^{2.62} & M < 0.7 M_\odot \\[2ex] 1.02 \left(\dfrac{M}{M_\odot} \right)^{3.92} & M > 0.7 M_\odot \end{cases} \tag{21.7}$$

and

$$\frac{R(M)}{R_\odot} = \begin{cases} 1.06 \left(\dfrac{M}{M_\odot} \right)^{0.945} & M < 1.66 M_\odot \\[2ex] 1.33 \left(\dfrac{M}{M_\odot} \right)^{0.555} & M > 1.66 M_\odot \end{cases} \tag{21.8}$$

Note that the mass appears in the functions normalized by the mass of the Sun $M_\odot$. These relations are an example of a particularly simple case of a fitting formula, called a "scaling relation", where the functional dependence is $f(x) = A(x/x_0)^\alpha$, that is, a dimensional form of the variable with some amplitude A and power law slope α.

The attentive reader will also note that Equations (21.7) and (21.8) have the slightly undesirable feature that they are not continuous at the $M = 0.7\ M_\odot$ for the luminosity and $M = 1.66\ M_\odot$ for the radius. (This requires plugging in and evaluating to confirm that the value from the two pieces is not, in fact, the same.) As explained in Demircan & Kahraman (1991), these actually represent the masses at which physically interesting changes take place within the stars.[3]

To implement the above relations as `Python` functions, we could use an `if` statement to select which power law to use based on the input mass. In the following implementation, we make the power law exponent and prefactor change smoothly between their two values over some relatively narrow range of stellar mass. This is not any more accurate than the above, but is aesthetically more pleasing (see Figure 21.2) and removes the discontinuity.

---

[3] In particular, this occurs when the kind of nuclear fusion in the core changes from the proton-proton fusion we're discussing here to other types.

```
def MSLuminosity(mass : u.kg) -> u.W:

    # express in terms of solar mass; ensure array
    functions behave properly by converting to an array
    m = (np.array(mass)*u.kg / c.M_sun).to(u.
    dimensionless_unscaled).value

    # a function to smoothly transition between the two
    power laws
    smooth_transition = 1/2*(np.tanh((m-0.7)/0.1)+1)

    exponent = 2.62 + smooth_transition*(3.92-2.62)
    prefac = (0.35 + smooth_transition*(1.02 - 0.35))*c.
    L_sun

    return prefac * np.power(m, exponent)

def MSRadius(mass : u.kg) -> u.m:

    # express in terms of solar mass; ensure array
    functions behave properly by converting to an array
    m = (np.array(mass)*u.kg / c.M_sun).to(u.
    dimensionless_unscaled).value

    smooth_transition = 1/2*(np.tanh((m-1.66)/0.1)+1)

    exponent = 0.945 + smooth_transition*(0.555 - 0.945)
    prefac = (1.06 + smooth_transition*(1.33 - 1.06))*c.
    R_sun

    return prefac * np.power(m, exponent)
```

Question 7. Use the functions MSLuminosity and MSRadius to make two plots, one of luminosity (in solar units) versus mass and the other of radius (again, in solar units) versus mass for masses between $0.1\ M_\odot$, and $50 M_\odot$. Make the plots log–log.

Question 8. Use MSLuminosity and MSRadius to construct another function MSTemperature and thus the ability to plot an H–R diagram as a function of stellar mass. Here the temperature will be in Kelvin. Calculate and plot L on the y-axis versus T on the x-axis from these relations for masses between $0.1 M_\odot$, and $50 M_\odot$. Again, log–log! Also, you can plot T increasing to left as is conventional by using plt.xlim([Tmax, Tmin]). Label your H–R diagram with the mass of the star for $M = 0.1, 1, 5, 50 M_\odot$.

21.5 Study Questions

1 **Information from the H–R Diagram.** Consider the H–R diagram in Figure 21.2 with 11 objects, which we will refer to as "stars" (even though not all are stars by the definition of actively fusing hydrogen to helium in their centers).

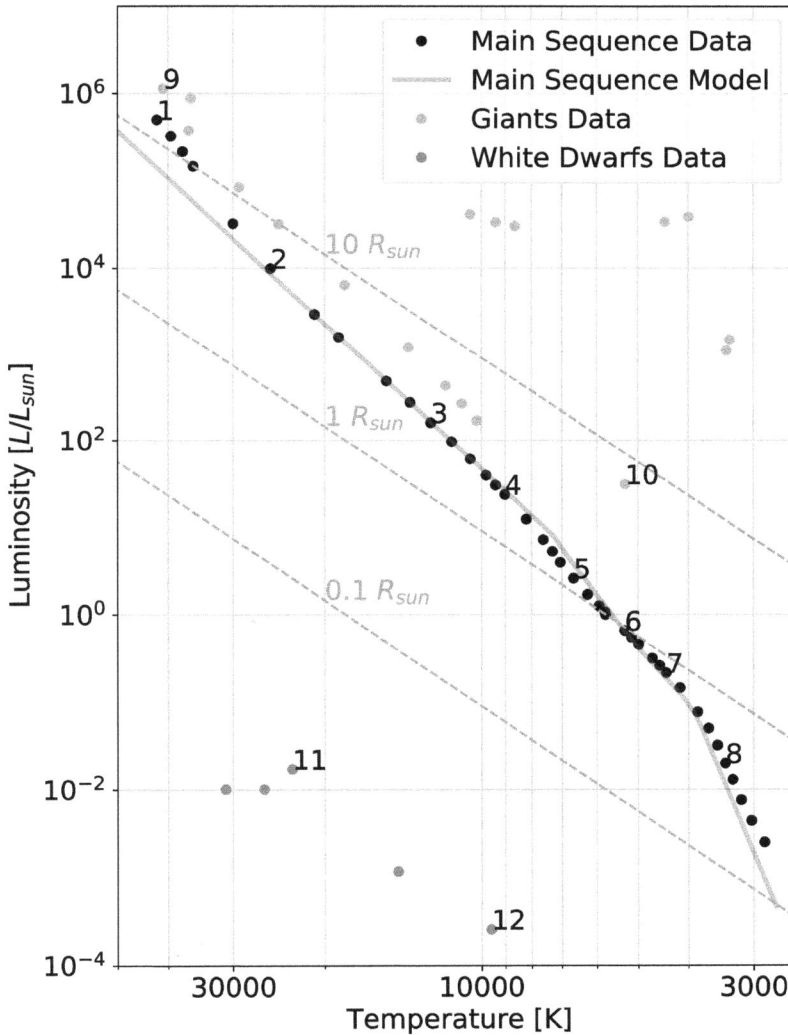

Figure 21.2. An H–R diagram showing three classes of objects: main sequence stars (black dots) with a model for the main sequence taken from Exercise III.6 (blue line). Giant stars (luminosity classes III and V) are shown in orange, and white dwarfs in maroon. Twelve objects are labeled, from 1 to 12. Note that lines of constant radius are marked at 0.1, 1, and 10 $R_{\odot}$; don't confuse these labels with the numbering of the objects. The data in this figure are taken from Carroll & Ostlie (2017).

(a) Which of the labeled stars has the largest radius? Which labeled star has the smallest radius? Estimate the ratio of their radii.

(b) If star 11 has a mass the same as our Sun, calculate its average density and compare that to the average density of the Sun.

(c) Which labeled star on the main sequence has the highest mass? Which labeled star has the lowest mass?

(d) ★ Estimate the mass of star 3 using one (or more) of the relationships between mass, luminosity, temperature, or radius.

(e) ★ Estimate the lifetime of star 3 on the main sequence, using the mass–luminosity relationship and either the fact that the Sun's lifetime is 10 billion years, or that 10% of the star's mass participates in nuclear fusion with an efficiency of 0.007. (You will get similar but not identical numbers for the two cases).

2 A certain main sequence star has a surface temperature $T = 6150$ K and luminosity of $L = 2L_\odot$ ($1L_\odot = 3.85 \times 10^{26}$ W). A planet moving around this star at a distance of 2 AU in a circular orbit causes a motion of the star around the center of mass with a maximum velocity of 7.5 m s^{-1}.

(a) What is the period of the planet's orbit around the star? (You will need to derive one piece of information not given in the statement of the problem to solve this.)

(b) Assume now that this planet actually transits its star as viewed by us. Using the fact (from momentum conservation) that $m_p = M_s(v_s / v_p)$ where m_p is the mass of the planet, M_s is the mass of the star, v_s is the velocity of the star around the system's center of mass, and v_p is the planet's velocity around its orbit, what is the mass of the planet?

(c) Does this planet receive a larger or smaller flux (in W m^{-2}) from its star than the Earth? Show your calculation.

References

Carroll, B. W., & Ostlie, D. A. 2017, An Introduction to Modern Astrophysics (2nd edn; Cambridge: Cambridge Univ. Press)

Demircan, O., & Kahraman, G. 1991, Ap&SS, 181, 313

An Introduction to Astrophysics with Python

Stars and planets

James Aguirre

Chapter 22

Absorption Lines in Stars

22.1 Stellar Atmospheres

We discuss how absorption lines form in stellar atmospheres and introduce the notion of line broadening. We also discuss the line cross section and derive a simple model for an absorption line using the radiative transfer equation. We bring together this model with the Boltzmann and Saha equations to make a qualitative model of the strength of absorption of the H-alpha line along the main sequence.

The "atmosphere" of a star (the part we see) is a very small part of the mass or even the energy budget of a star. But because we can see it, it has the bulk of the *information* about the star. In looking at a stellar atmosphere, we can determine, among other things, the surface temperature, the composition (what elements are present), the density of the material, the strength of gravity at the surface of the star (related to its mass and radius), information about the rotation of the star, and finally information about the relative motion of the star toward or away from us (using the Doppler effect, as in Chapter 16). All this interesting information comes from the *line* transitions of atoms and ions, since the *continuum* emission has very few features (in the ideal case, the *blackbody* continuum emission only tells us about the temperature of the object, but nothing about its composition, etc.). The depth and width of the lines has a very rich structure, however.

Let's think about the outer atmosphere of a star made of pure hydrogen. If the temperature is low enough, we expect that all the hydrogen is in the ground state, and there's not much probability of collisional excitation, and thus we won't see line emission or absorption, because transitions to any excited states are rare. If the temperature is too high, all the hydrogen atoms will be ionized, and we won't see any line emission or absorption in that case, either. This suggests there is an interesting

doi:10.1088/2514-3433/ade5f6ch22   22-1

range of temperatures where we might expect to see some non-negligible fraction hydrogen atoms in excited states where they can absorb and emit line emission.

The "big ideas" of this section are understanding what information we can get from looking at spectral lines, and in particular at their width and depth (or height). We're going to think about absorption lines in stellar atmospheres, but it could apply just as well to emission lines, since the emission coefficient is responsible for both how strong the emission line is and how strong the absorption.

22.2 Line Width

In discussing line emission and absorption in atoms in Chapter 12, we noted that there are definite wavelengths absorbed or emitted due to the discrete energy levels of the atom, and we said that the emission or absorption of a photon had to be *exactly* this wavelength to interact with the atom. Of course, nothing is quite exact, and we now want to ask, what mechanisms are there that might allow an atom to absorb or emit at wavelengths different from the precise energy difference between levels? This range of wavelengths or frequencies over which the atom (or a group of atoms) might be able to absorb or emit we'll refer to as the "line width". Much like the probability of an atom having a particular velocity described by the Maxwell–Boltzmann distribution, we're going to use the idea of a probability distribution to describe how likely it is that an atom will interact with a given frequency of light, given by a function $\phi(\nu)$. The probability density $\phi(\nu)$ will be small (close to zero) for frequencies far from the allowed ones and only large for frequencies close to the "exact" frequency corresponding to the energy difference of the transitions.

While there are a number of mechanisms that can affect the line width, we're going to consider just two important ones:

- the"natural" line width. This is a fundamental feature of the behavior of atoms because of quantum mechanical uncertainty and because of the inability to precisely determine an energy in a finite amount of time.
- "thermal Doppler broadening". This is not a property of an individual atom, but rather a feature of the collective motion of many atoms in thermal motion: the Doppler shift causes some of them to emit or absorb at longer or shorter wavelengths than they would if they were at rest. An observer at rest relative to this collection of atoms will see that interaction with the line occurs over a range of wavelengths.

A real spectral line will have both natural and Doppler broadening, the combined effect of which is called a "Voigt profile", for which a simple formula cannot be written down. However, there is an astropy function that will calculate it for you.

We now consider the quantitative description of the natural and Doppler profiles. The natural line width of a line is given by the so-called Lorentz distribution

$$\phi_L(\nu)d\nu = \left(\frac{\gamma_n/4\pi}{(\nu - \nu_0)^2 + (\gamma_n/4\pi)^2} \right) \frac{d\nu}{\pi} \tag{22.1}$$

where the line frequency ν_0 is defined from the energy change $\Delta E = h\nu_0$—that is, is the frequency we would calculate from the energy levels. The constant γ_n is the *damping constant*, which describes how long it takes the atom to make this transition, and is specific to each transition the atom can make. With some calculus, we can show that $\phi_L(\nu)$ is a properly normalized probability distribution, that is,

$$\int_{-\infty}^{\infty} \phi_L(\nu)d\nu = 1 \qquad (22.2)$$

To get an expression for the thermal Doppler broadening for an ensemble of atoms with temperature T and particle mass μm_p (m_p= mass of the proton), we first need to calculate the Doppler shift in frequency, which will depend on the atom's velocity, and then combine this with the Maxwell–Boltzmann distribution for the probability of an atom having a given velocity.

To find the Doppler shift in terms of frequency, we start by using the relation between frequency and wavelength $c = \lambda\nu$, to find that

$$\frac{d\nu}{d\lambda} = -\frac{c}{\lambda^2} \qquad (22.3)$$

This means that small changes in frequency are related to small changes in wavelength by

$$\frac{d\nu}{\nu} = -\frac{1}{\nu}\frac{c}{\lambda^2}d\lambda = -\frac{d\lambda}{\lambda} \qquad (22.4)$$

Thus, for small fractional changes in wavelength or frequency, we can write the Doppler shift formula as

$$\frac{v_r}{c} = \frac{\lambda - \lambda_0}{\lambda_0} = \frac{\nu_0 - \nu}{\nu_0} \qquad (22.5)$$

Now, we want to combine this shift in the wavelength or frequency with the distribution of velocities we can get from the *one-dimensional* version of the Maxwell–Boltzmann distribution (since we only care about line-of-sight velocities). This is given by a Gaussian distribution in the line-of-sight velocity v_r:

$$\phi(v_r)dv_r = \left(\frac{1}{2\pi\sigma_v^2}\right)^{1/2} \exp\left(-\frac{v_r^2}{2\sigma_v^2}\right)dv_r \qquad (22.6)$$

where

$$\sigma_v^2 = \frac{kT}{\mu m_p} \qquad (22.7)$$

This is actually the same σ_v as for the 3D case.

To transform this from a probability of an atom having a given velocity to the probability of an atom interacting with a shifted frequency of light due to its Doppler shift, we need to write the velocity in terms of the associated shifted frequency:

$$v_r = c\frac{(\nu_0 - \nu)}{\nu_0} \tag{22.8}$$

and

$$dv_r = -c\frac{d\nu}{\nu_0} \tag{22.9}$$

Substituting these into Equation (22.6), we get

$$\phi(v_r)dv_r = \phi(\nu)d\nu = \left(\frac{1}{2\pi\sigma_v^2}\right)^{1/2} \exp\left(-\left(\frac{c(\nu - \nu_0)}{\nu_0}\right)^2 \frac{1}{2\sigma_v^2}\right)\frac{c}{\nu_0}d\nu \tag{22.10}$$

$$= \frac{c}{\nu_0}\left(\frac{1}{2\pi\sigma_v^2}\right)^{1/2} \exp\left(-\frac{1}{2}\left(\frac{c(\nu - \nu_0)}{\nu_0\sigma_v}\right)^2\right) \tag{22.11}$$

$$= \left(\frac{1}{2\pi\sigma_\nu^2}\right)^{1/2} \exp\left(-\frac{1}{2}\left(\frac{(\nu - \nu_0)}{\sigma_\nu}\right)^2\right) \tag{22.12}$$

where we've defined a new width[1] of the line in *frequency* by[2]

$$\sigma_\nu = \nu_0\frac{\sigma_v}{c} = \frac{\nu_0}{c}\sqrt{\frac{kT}{\mu m_{\rm p}}} \tag{22.13}$$

Writing it all out explicitly, we get the

$$\phi_D(\nu)d\nu = \frac{c}{\nu_0}\sqrt{\frac{\mu m_{\rm p}}{2\pi kT}} \exp\left[-\frac{\mu m_{\rm p}c^2(\nu - \nu_0)^2}{2kT\nu_0^2}\right]d\nu \tag{22.14}$$

22.3 The Line Cross Section

The details of how we can calculate the cross-section for a photon of given energy interacting with a particular energy transition in an atom are fairly complicated, and require a full quantum mechanical treatment of the atom. We won't go into these details, because the result turns out to have a simple form. This absorption cross-section for a photon interacting with a given line transition with profile $\phi(\nu)$ is

$$\sigma(\nu) = \frac{e^2}{4\epsilon_0 m_{\rm e}c}f\phi(\nu) \tag{22.15}$$

where $e = 1.6022 \times 10^{-19}$ C is the fundamental electrical charge, and $\epsilon_0 = 8.8542 \times 10^{-12}$ C^2 N^{-1} m^{-2} is the constant appearing in the Coulomb force

---

[1] Here we're taking the width to be the standard deviation of the Gaussian distribution.
[2] Note the potential confusion between v, the velocity, and ν, the frequency.

law, and f is a dimensionless quantity called the *oscillator strength*, which is different for each different atomic transitions in an atom. *All of the messy quantum mechanics has been absorbed in the single, dimensionless number f.* The cross-section $\sigma(\nu)$ has units of area (m^2). Then we can write

$$\tau(\nu) = \sigma(\nu)nx = \sigma(\nu)N_a \tag{22.16}$$

where N_a is the *column density* of the atoms.

From our previous discussion of photons traversing a slab of thickness x, we can then write the *spectrum* of the star near the absorption line using our simple absorption model that says the flux F after passing through a slab of thickness x with number density n and cross-section for interaction σ is given by

$$F = F_0 \exp(-\sigma nx) \tag{22.17}$$

where F_0 was the original flux incident on the slab. We usually write the spectrum normalized to the incident (continuum) flux to just show how the spectrum is attenuated near the line:

$$\frac{I_{\text{line}}}{I_{\text{continuum}}} = \exp\left[-\tau(\nu)\right] \tag{22.18}$$

This somewhat misleadingly simple equation gives a model for an absorption line. As we explore further in the exercise, it does not properly capture all the features of real absorption lines, and also a lot of detail is hidden inside the calculation of the function $\tau(\nu)$.

22.4 Exercise: Modeling an Absorption Line

Goals for this exercise:
   1. Understand the ingredients in the formation of an absorption line

Here we will work out the behavior of the $n = 3 \to 2$ absorption line in the atmosphere of the Sun. We take the temperature of the stellar atmosphere to be $T = 5800$ K and its density to be 1×10^6 kg m^{-3}, and for simplicity, we will assume it is all hydrogen (so that $\mu = 1$). If we zoom in on a portion of the solar spectrum near the $n = 3 \to n = 2$ hydrogen transition, which astronomers call Hα, it looks like Figure 22.1.

There are clearly a lot of absorption lines present (all the little notches and divots), but let's focus on the biggest one, Hα. We are going to try to match some general features of this absorption line and extract some information about how the line was formed. The broad "wings" that you see are actually due to so-called "pressure broadening", which mathematically we can describe with the Lorentz profile, just like natural broadening, but with a different γ_n.

Question 1. What is the number density of hydrogen atoms in the stellar atmosphere?

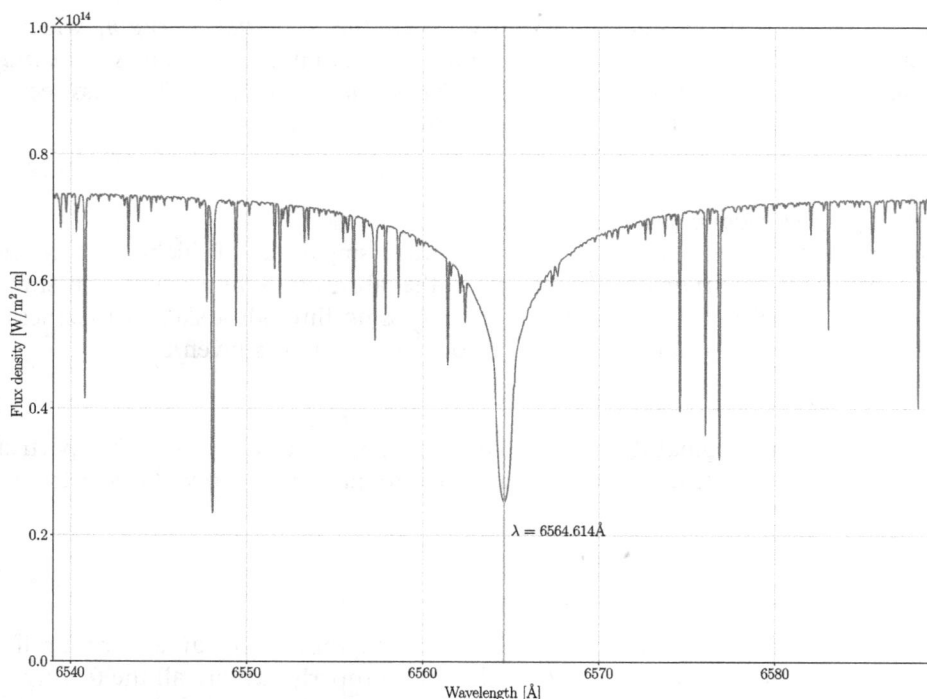

Figure 22.1. A portion of a simulated stellar atmosphere spectrum with $T_{\text{eff}} = 5800$ K and mass and radius similar to the Sun, for wavelengths near the Hα line. This is from the library of stellar spectra described in Husser et al. (2013). Note that 1 Angstrom $= 1 \times 10 - 10$ m.

In the exercise in Chapter 15, we used the Saha equation to work out the fraction of ionized H atoms in the Sun's atmosphere. We can combine that with the Boltzmann equation for the neutral hydrogen atom to get the fraction of H atoms in the $n = 2$ state.

Question 2. Under the approximation that $n_{\text{HI}} = n_1 + n_2$ (no atoms are in the $n = 3$ or higher states), write a function that will calculate the fraction of H atoms that are in the $n = 2$ state at a given temperature (which must have units of K). Plot this function for the temperature range $5000 \leqslant T \leqslant 25,000$ K. At what temperature is the largest fraction of atoms in the $n = 2$ state? Approximately what is that fraction?

The answer to Question 2 makes a strong prediction: for stars of that temperature, we will see strong *absorption* for the red H-α $n = 3 \rightarrow 2$ line, because photons can be absorbed by the electrons in the $n = 2$ state, rise up to the $n = 3$ state, and then be re-emitted when they fall back down to $n = 2$ or $n = 1$ into some direction not where they were originally traveling, thus being lost back into the sea of photons in the stellar atmosphere.

Question 3. Using the results of Questions 1 and 2, at 5800 K, what is the number density of $n = 2$ atoms?

These are the atoms that will participate in the formation of the absorption line. Let's now try to calculate the observed spectrum of this line.

The Doppler line width function (Equation 22.14), which dominates the line width, and so is sufficient for use here, is implemented in the code below.

```
@u.quantity_input
def SigmaV(T : u.K,mu) -> (u.m/u.s):

    m = mu*c.m_p
    sigma_v = np.sqrt(c.k_B * T / m)
    return sigma_v

@u.quantity_input
def Doppler(nu : u.Hz, nu0: u.Hz, T: u.K, mu) -> u.Hz**-1:

    s = SigmaV(T, mu)

    x = nu - nu0

    exparg = (- np.power(x * c.c / ( nu0 * s), 2)/2.).to(u.
    dimensionless_unscaled)

    phi = c.c/nu0 * np.power(np.sqrt(2*np.pi) * s, -1) * np
    .exp(exparg)

    return phi
```

Question 4. Let's consider the transition $n = 3 \rightarrow n = 2$ in the Sun. The damping constant for this line is $\gamma_3 = 4.4 \times 10^7$ Hz. Make a plot which has the natural and Doppler profiles overplotted for this same line. At about what values of $\xi = (\nu - \nu_0)/\nu_0$ does each distribution reach half its maximum value? (You can read this off approximately from the plot or the calculated function values.) Note that the natural width is *very* narrow and the peaks of the distributions are very different (since they are normalized to all have an area of 1). To put them on the same plot, it will be best to make the peak heights equal to 1, and plot the y-axis logarithmically. You will need to play with the scales to get this to look nice. Note that $|\xi| < 0.01$ will capture most of the interesting features.

To finish calculating the spectrum of the absorption line, we need to know the cross-section for photon interaction with a hydrogen atom in the $n = 2$ state, which we look up, and at which line center we will take to be $\sigma_0 = 5 \times 10^{-20}$ m^2. The last thing we need is the thickness of the atmosphere, and we'll use $\ell = 500$ km.

Question 5. Using the given cross-section and thickness, and the number density from Question 3, calculate the opacity at line center τ_0. Do we expect the line to be very strong (i.e., a lot of absorption)?

Question 6. Plot the *spectrum* of the star near the absorption line, which is given by

$$\frac{I_{\text{line}}}{I_{\text{continuum}}} = \exp[-\tau_0 \Phi(\nu)] \qquad (22.19)$$

where you can assume that

$$\Phi(\nu) = \frac{\phi(\nu)_{\text{Doppler}}}{\phi(\nu_0)_{\text{Doppler}}} \qquad (22.20)$$

and, near the line, the continuum is flat.

Reference

Husser, T. O., Wende-von Berg, S., Dreizler, S., et al. 2013, A&A, 553, A6

AAS | IOP Astronomy

An Introduction to Astrophysics with Python
Stars and planets
James Aguirre

Chapter 23

A Model of Stellar Structure

Building on the discussion in Chapter 20, which introduced the differential equations describing pressure and mass distributions in spherically symmetric objects, we add the additional equations describing energy generation via nuclear fusion the flow of heat via a temperature gradient in the radiation field to produce a complete-enough model for the structure of stellar interiors. We examine the behavior of the solutions of these equations for a model of the Sun taken from the literature.

Our goal in this chapter is to derive and understand the four equations of *stellar structure*. In the section on hydrostatic equilibrium, we found that for a spherical, self-gravitating[1] object (like the Sun) to be stable, it was necessary to balance the inward pull of gravity with a pressure gradient, generated in some way. This led to two equations, the first defining the relation between the mass enclosed in a radius $M_r(r)$ and the density as a function of radius $\rho(r)$

$$\frac{dM_r(r)}{dr} = 4\pi r^2 \rho(r) \qquad (23.1)$$

and the second defining the required pressure gradient. Notice that the pressure gradient is always negative, that is, decreasing in pressure from the center out:

$$\frac{dP(r)}{dr} = -G\frac{M_r(r)}{r^2}\rho(r) \qquad (23.2)$$

---

[1] Self-gravitating here just means that the gravitational force dominates in describing the structure of the object. You are not self-gravitating, because the gravitational pull of various parts of your body on each other are vastly smaller than the electromagnetic forces between them.

doi:10.1088/2514-3433/ade5f6ch23

Notice our equations express $P(r)$, $\rho(r)$, and $M_r(r)$ as functions of the distance r from the center of the star.[2] Since we cannot directly observe the interior of the star, these equations are going to be our guide to the conditions there, and we would have a pretty good understanding of the interior if we had these functions.

However, there are already a couple of problems in solving these equations. We have two differential equations, one for M_r and one for P, but these both depend on ρ, which is generally unknown to start.[3] Even if we are able to specify $\rho(r)$, we also need initial conditions: $M_r(0)$ and $P(0)$. For $M_r(0)$ we can work out the initial condition, since it is, by definition, the mass enclosed in a radius of 0: that must be zero. However, we don't know the central pressure $P(0)$ - this is one of the things we're trying to find. However, the "initial" conditions don't have to be specified at $r = 0$. If we know the total radius R and the total mass M, we can also start the integration at the surface and work inward: $M_r(R) = M$ and $P(R) = P_{surf}$, which at least has the virtue of being an observable quantity, and often P_{surf} is chosen to be 0. This is actually what we did in Chapter 20: assume $\rho = \rho_0$ and integrate Equation (23.1) from 0 to R to get $M_r(r)$, and then insert $M_r(r)$ into Equation (23.2) and integrate, using the fact that $P(R) = 0$.

We can also note that, on physical grounds, there is usually some sort of relation between pressure, temperature, and density, expressed as

$$P = f(\rho, T) \tag{23.3}$$

For example, we have the ideal gas law which relates density, pressure, and temperature:

$$P = \frac{\rho}{\bar{m}} kT \tag{23.4}$$

In addition, there may be pressure due to radiation and other quantum mechanical effects. Although it may be surprising, the ideal gas law is still a pretty good approximation for much of the interior of the sun. This *constituent relation* between T, ρ and P provides a crucial link between these quantities and motivates us to also have some equation governing $T(r)$.

Clearly, the differential equation for T will be related to how energy is flowing in the star. We expect that the inner regions of the sun will be hotter and the outer ones cooler; moreover, we know that heat must flow from hot to cold, so the direction of temperature change dT/dr must be negative so that heat flows from hot to cold. We know that there are three methods of transferring heat: conduction, convection, radiation, and we can largely ignore conduction and (at least to a first approximation) ignore convection for a star like the sun. So we first want to know about energy transport through radiation. But before we talk about energy *transport*, and what an equation for dT/dr might look like, we need to talk about energy *generation*.

---

[2] For brevity, this dependence will sometimes be dropped, but always remember that these are functions.
[3] We did work out a solution in Chapter 20 for the case that $\rho(r) = \rho_0 =$ constant, appropriate for matter that is incompressible (density doesn't depend on pressure) and whose density also does not depend strongly on temperature.

23.1 Nuclear Fusion in Stars

Here we will not attempt a comprehensive discussion of nuclear physics in stars but will only attempt to understand some of the key processes at play and how we can describe them phenomenologically. In building a model of stellar structure, which is our goal in the last portion of this book, we do not need to track individual nuclear reactions but rather seek to describe the dependence of the nuclear fusion rate on the density and temperature at some point inside the star. We note that the amount of energy generated via fusion is a strong function of the temperature (to overcome the Coulomb barrier between fusing nuclei), and also depends on the density of the material and the composition (and in particular what kind of nuclear fusion processes are possible). Without getting into the details of the nuclear physics, we can say that the energy generation rate from the fusion of 4 protons into a helium nucleus, from which the Sun derives most of its power, increases as the density and temperature of the material increases: increasing the density puts more nuclei in closer proximity, and increasing the temperature means that they have more kinetic energy to overcome their electrostatic repulsion (since the protons colliding all have positive charge). We can write down a phenomenological expression for this called the luminosity per unit mass

$$\epsilon_{pp} = (1.08 \times 10^{-12} \text{ Wm}^3 \text{ kg}^{-2}) X^2 \rho \left(\frac{T}{10^6 \text{ K}} \right)^4 \qquad (23.5)$$

where the units of ϵ_{pp} are W kg^{-1}. This function depends on the mass fraction of hydrogen X and the density and temperature of the material, and increases fairly steeply as the temperature goes up.

With this description in hand, we can write the change in luminosity for the addition of a little luminosity for a little mass as

$$dL_r = \epsilon dm = \epsilon(T, \rho)\rho dV = \epsilon\rho 4\pi r^2 dr \qquad (23.6)$$

so that we now have our next stellar structure equation

$$\frac{dL_r}{dr} = 4\pi r^2 \epsilon(T, \rho)\rho(r) \qquad (23.7)$$

Now, the function L_r is relatively easy to grasp: if the material at radius r is fusing hydrogen into helium, then ϵ is non-zero, and the material is releasing energy and thus generating luminosity; otherwise, the material is simply passing the radiation through it, and L_r measures the total amount of luminosity generated interior to r.

Finally, we'd like our model to respect one of the first assumptions of our simple stellar model, namely that

$$L_* = 4\pi R_*^2 \sigma T_*^4 \qquad (23.8)$$

As we have seen, this is a simplification, and the only really well-defined quantity in this equation is the bolometric luminosity L_*, which for a main-sequence star is fairly constant. The radius R is ill-defined because, of course, there is no surface, and the stellar atmosphere doesn't have a well-defined boundary. So we are going to have to pick what we mean by the stellar radius. There are two main criteria we want a definition of stellar radius to obey, the first being that the radius enclose, to a very good approximation, all of the stellar mass, or

$$M_* = 4\pi \int_0^{R_*} \rho(r) r^2 dr \qquad (23.9)$$

This neglects any mass in the atmosphere where $r > R_*$ In order that (23.8) is satisfied, we're also going to have to assert that the radius R_* corresponds to the height in the stellar atmosphere at which the temperature is equal to the effective temperature T_*, that is

$$T(r = R_*) = T_* \qquad (23.10)$$

23.2 Flux due to a Temperature Gradient

We know that radiation can transfer energy through a star. Let's derive an equation for the transport of energy by radiation where, at least locally, there is a concept of temperature, and the radiation field is locally described by a blackbody. The radiation pressure of blackbody radiation is given by

$$P_{\text{rad}} = \frac{4\sigma}{3c} T^4 \qquad (23.11)$$

where σ is the Stefan–Boltzmann constant appearing in the equation

$$F_{\text{surf}} = \sigma T^4 \qquad (23.12)$$

for the flux of radiation at the surface of a blackbody of temperature T.

Even though the radiation pressure doesn't dominate the total pressure, it is responsible for the movement of the heat inside the Sun. Let's see how this comes about. We can start to see where an equation for dT/dr comes from by differentiating Equation (23.11) with respect to r:

$$\frac{dP_{\text{rad}}}{dr} = \frac{16\sigma}{3c} T^3 \frac{dT}{dr} \qquad (23.13)$$

However, we still don't know how left-hand side of this equation is related to other quantities in the problem. For that, we need to remember that we defined the differential change in the optical depth τ_λ of a material as

$$d\tau_\lambda = n\sigma_\lambda dz = \kappa_\lambda \rho dz \qquad (23.14)$$

where the λ subscript reminds us that this is the optical depth (or cross-section, or opacity) to photons of wavelength λ.

The appearance of the optical depth in the problem shouldn't come as a surprise: the transfer of energy is clearly easier if the material is more transparent, and harder if it is more opaque. An important result that follows from the application of the radiative transfer equation is that

$$\frac{dP_{rad}}{dr} = -\frac{\bar{\kappa}\rho}{c}F_{rad} \tag{23.15}$$

where F_{rad} is the flux of the radiation (integrated over all wavelengths) and $\bar{\kappa}$ is the opacity averaged over wavelengths.

Now suppose that at some radius r, there is a total luminosity of radiation L_r that has been generated (by nuclear fusion) interior to that radius. It is relatively easy to see that, from our definition of flux, the flux passing through an imaginary surface at radius r inside the star must be

$$F_{rad} = \frac{L_r}{4\pi r^2} \tag{23.16}$$

So now we can write

$$\frac{dP_{rad}}{dr} = -\frac{\bar{\kappa}\rho}{c}\frac{L_r}{4\pi r^2} \tag{23.17}$$

and

$$\frac{dP_{rad}}{dr} = \frac{16\sigma}{3c}T^3\frac{dT}{dr} \tag{23.18}$$

Setting these equal gives

$$\frac{dT}{dr} = -\frac{3\bar{\kappa}\rho L_r}{64\pi\sigma r^2 T^3} \tag{23.19}$$

This gives us the relation we need for the transport of energy via radiation.

23.3 The Stellar Structure Equations

We have learned a number of things about the structure of stars and their properties. In this concluding section, we'd like to put together a model of a star that encapsulates the things we've learned, is internally self-consistent, and captures most of the interesting features of a stable, main-sequence star near the beginning of its life, the so-called zero-age main sequence or ZAMS.

The result of considering hydrostatic equilibrium (balance between gravitational attraction and a pressure gradient), as well as energy generation (from nuclear fusion) and transport (largely due to radiation) in a star's interior leads to four, first-order differential equations for the *stellar structure*. These four equations are

$$\frac{dM_r}{dr} = 4\pi r^2 \rho(r) \tag{23.20}$$

$$\frac{dP}{dr} = -G\frac{M_r(r)\rho(r)}{r^2} \tag{23.21}$$

$$\frac{dL_r}{dr} = 4\pi r^2 \rho(r)\varepsilon[T(r),\, \rho(r),\, X] \tag{23.22}$$

$$\frac{dT}{dr} = -\frac{3}{4ac}\frac{\kappa[T(r),\, \rho(r),\, X]\rho(r)}{T^3(r)}\frac{L_r(r)}{4\pi r^2} \tag{23.23}$$

where we have noted all the explicit dependencies of the variables on the independent variable r. The quantity X denotes the composition (that is, the mass fractions of hydrogen, helium, and other elements). It could also be a function of position, but we will not worry about this complication.

We also have some *constituent equations* that are functions of $\rho(r)$ and $T(r)$ that give us

$$P = P[\rho(r),\, T(r)] \tag{23.24}$$

$$\kappa = \kappa[\rho(r),\, T(r)] \tag{23.25}$$

$$\varepsilon = \varepsilon(\rho(r),\, T(r)) \tag{23.26}$$

The pressure can be written as the sum of the ideal gas and radiation pressure

$$P = \frac{\rho k T}{\mu m_H} + \frac{1}{3}aT^4 \tag{23.27}$$

The approximate opacity is given by

$$\kappa = (\kappa_{bf} + \kappa_{ff})\rho T^{-3.5} + \kappa_{es}(1 + X) \tag{23.28}$$

and the energy generation for the various nuclear processes as

$$\epsilon = \rho^n T^m \tag{23.29}$$

with the proton–proton chain, for example, given by Equation (23.5).

23.4 The Structure of the Sun

In this final section, we consider the application of these equations to the internal structure of the Sun, using the model presented in Bahcall et al. (2001), which is more sophisticated than our simple model above; in particular, it accounts for convection as well as changes in elemental composition as a function of radius. In all Figures 23.1 to 23.11, we have plotted each of the quantities of interest versus the radius (in units of the stellar radius), linearly in the quantity in the top panel, and logarithmically in the bottom, to better appreciate the large dynamic range in these quantities. We have also color-coded three zones in the Sun: the inner portion, or *core* is where most of the nuclear fusion occurs; the *radiation zone*, where radiation dominates the transport of energy outward; and the *convection zone* near the surface, where convection takes over and dominates the energy transport. Note that this ordering of the zones does not apply to all stars on the main sequence.

We first examine the density $\rho(r)$ and enclosed mass $M_r(r)$ as a function of the radius in Figures 23.1 and 23.2, respectively. The key conclusion is that the mass is strongly concentrated towards the center.

Turning now to the pressure, the solution to Equation (23.20) is shown in Figure 23.3. Skipping slightly ahead, we can also compute the ideal gas pressure from the solutions for ρ and T, and compare it to the total pressure; this is shown in Figure 23.4. Note that the total pressure is indeed dominated by the ideal gas pressure (and not by radiation or other effects), at least for the Sun.

The luminosity is primarily generated in the hottest, densest central regions. Figure 23.5 shows the enclosed luminosity as a function of radius, and Figure 23.6

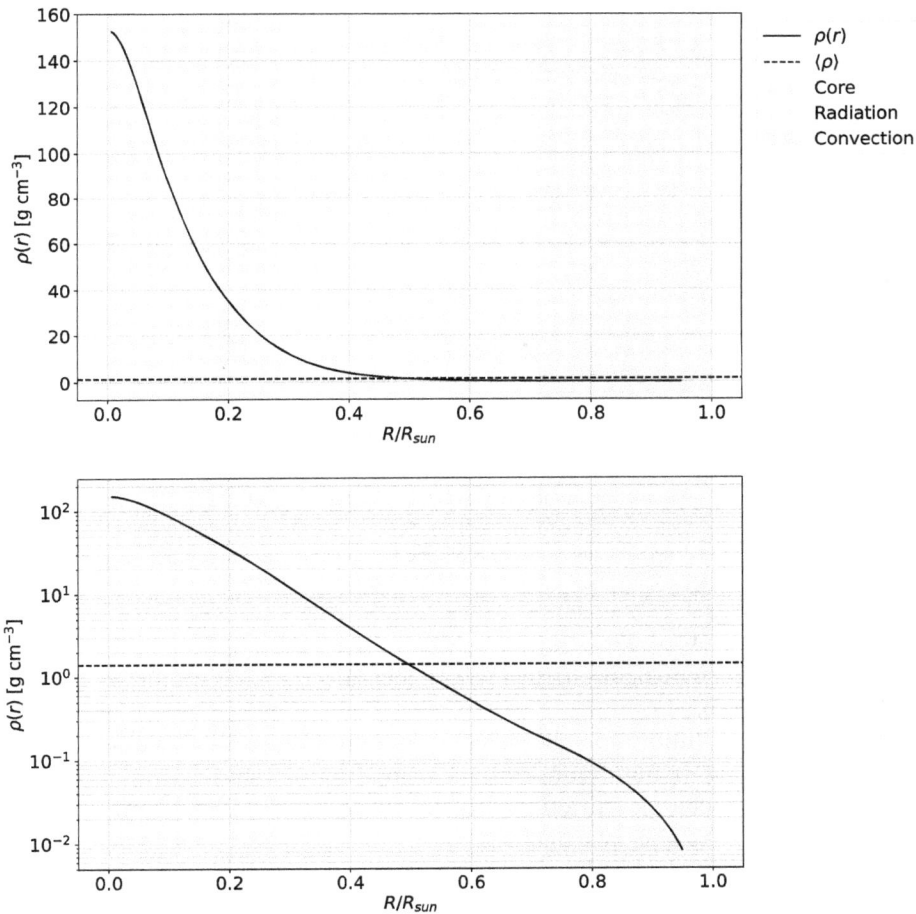

Figure 23.1. Density as function of radius $\rho(r)$. This is not directly the result of solving any of the stellar structure equations, but is the density that is consistent with the solution of Equation (23.21). Note the steep fall-off of density with radius.

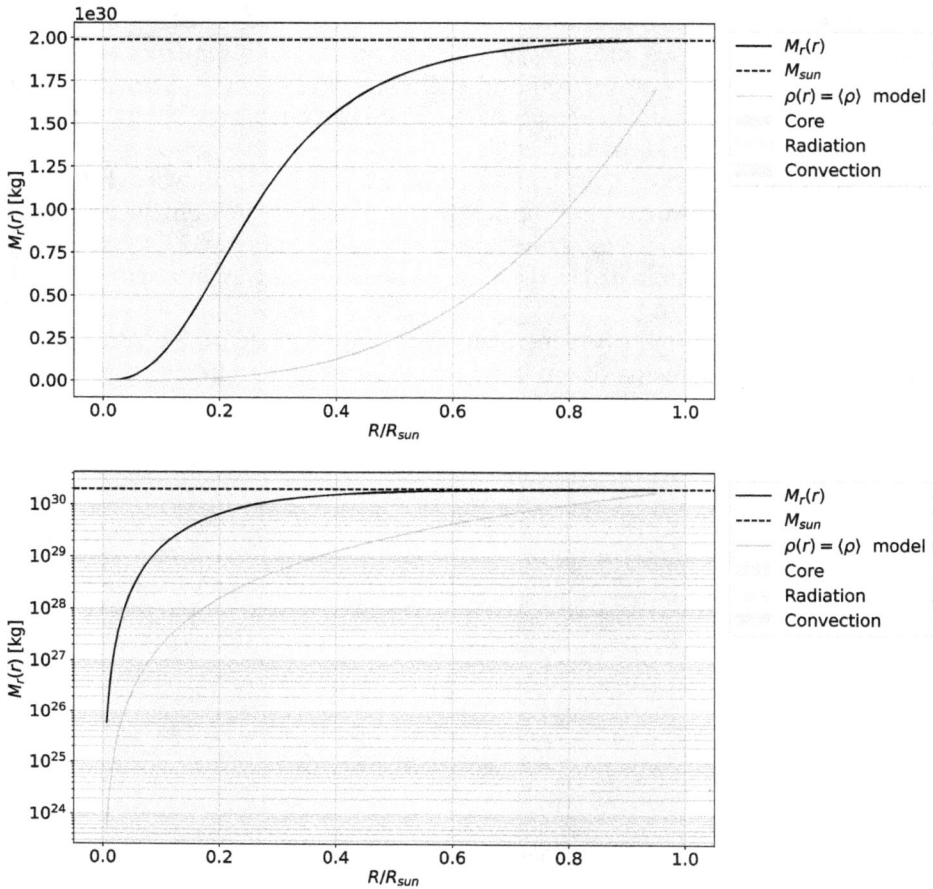

Figure 23.2. Enclosed mass as a function of radius. This is the solution to Equation (23.21). Viewed this way, it is clear that about half of the mass is in the core, at radii $r < 0.25 R_\odot$, which also represents a few percent of the total volume of the Sun. The constant density model is plotted for comparison, where most of the mass is in the outer regions.

shows the power from nuclear energy [W/kg] generated at each radius, as computed from Equation (23.5).

Examining the solution for the temperature in Figure 23.7 we see that the temperature gradient is quite constant (in a logarithmic sense) over most of the stellar interior but changes dramatically in the outer region where convection actually takes over the energy transport.

Not considered in our model, but of great practical importance in real stars is the change in composition through the star as nuclear fusion proceeds and as convection mixes up the internal material. Figure 23.8 shows the model composition versus radius, dominated by hydrogen and helium.

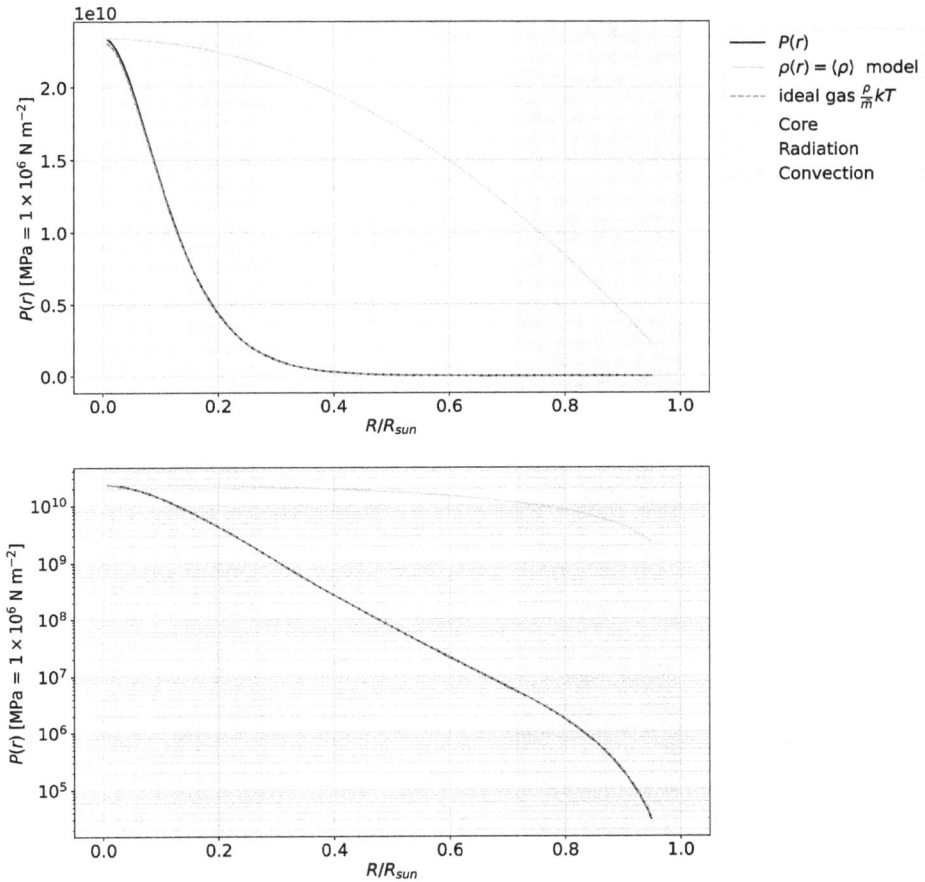

Figure 23.3. Pressure as a function of radius. This is the solution to Equation (23.20). The pressure implied in the constant density model is also plotted. Note the central concentration of high pressures, but also the monotonic decrease required so that dP/dr is always negative.

Examining the solution for the temperature in Figure 23.7, we see that the temperature gradient is quite constant (in a logarithmic sense) over most of the stellar interior but changes dramatically in the outer region where convection actually takes over the energy transport.

Not considered in our model, but of great practical importance in real stars is the change in composition through the star as nuclear fusion proceeds and as convection mixes up the internal material. Figure 23.8 shows the model composition versus radius, dominated by hydrogen and helium.

Finally, we conclude by comparing the left- and right-hand sides of the stellar structure equations by direct numerical differentiation (recalling Chapter 7) to see how well the derivatives of M_r (Figure 23.9), P (Figure 23.10) and L_r (Figure 23.11) from the Bahcall et al. (2001) model compare against the right-hand sides of

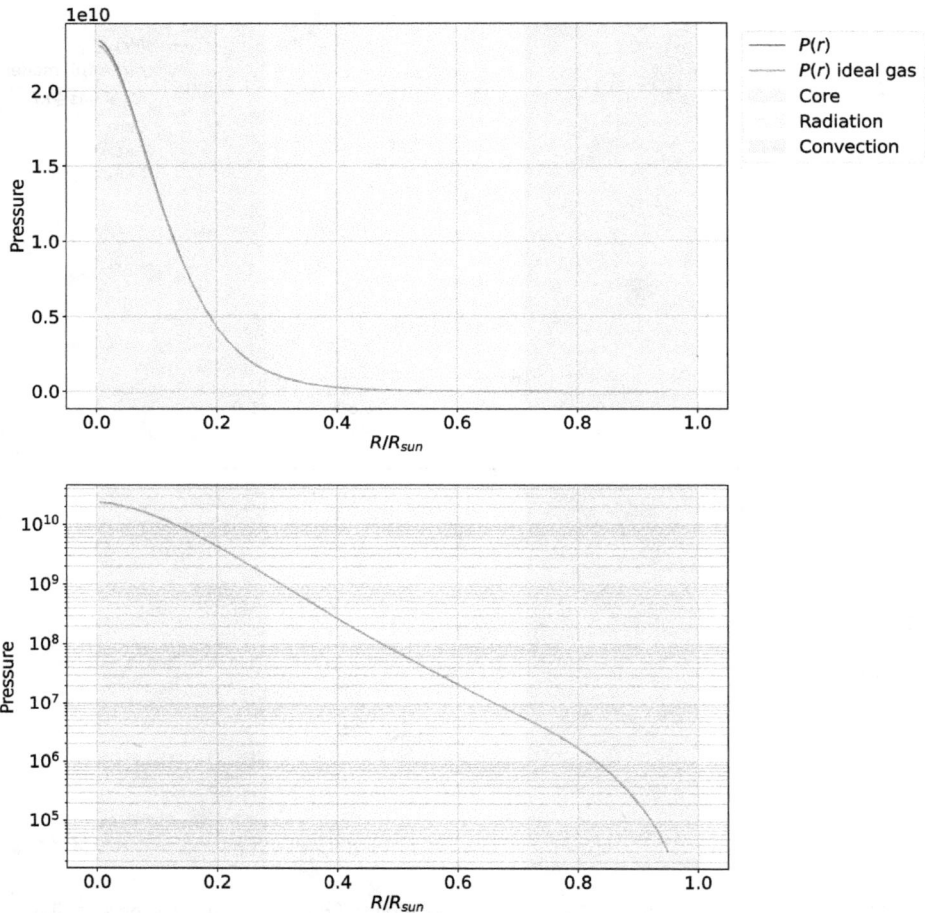

Figure 23.4. Comparison of the model pressure P to that of an ideal gas with the density ρ and temperature T computed from the model. The two quantities agree too well to be told apart on the plot.

Equations 23.21, 23.20, and 23.22, respectively. You will note that the numerical derivatives show some "noise" due to the finite step size, but the agreement is quite good. Of particular note is that in Figure 23.11, if we compute the right-hand side using Equation (23.5), we get a fairly good match, *until* we reach the edge of the core, where the Bahcall et al. (2001) model has the energy generation drop effectively to zero, but our simple model continues to predict fusion. We note that if we were to follow this procedure for Equation (23.23), we would be able to infer the opacity κ used in the model.

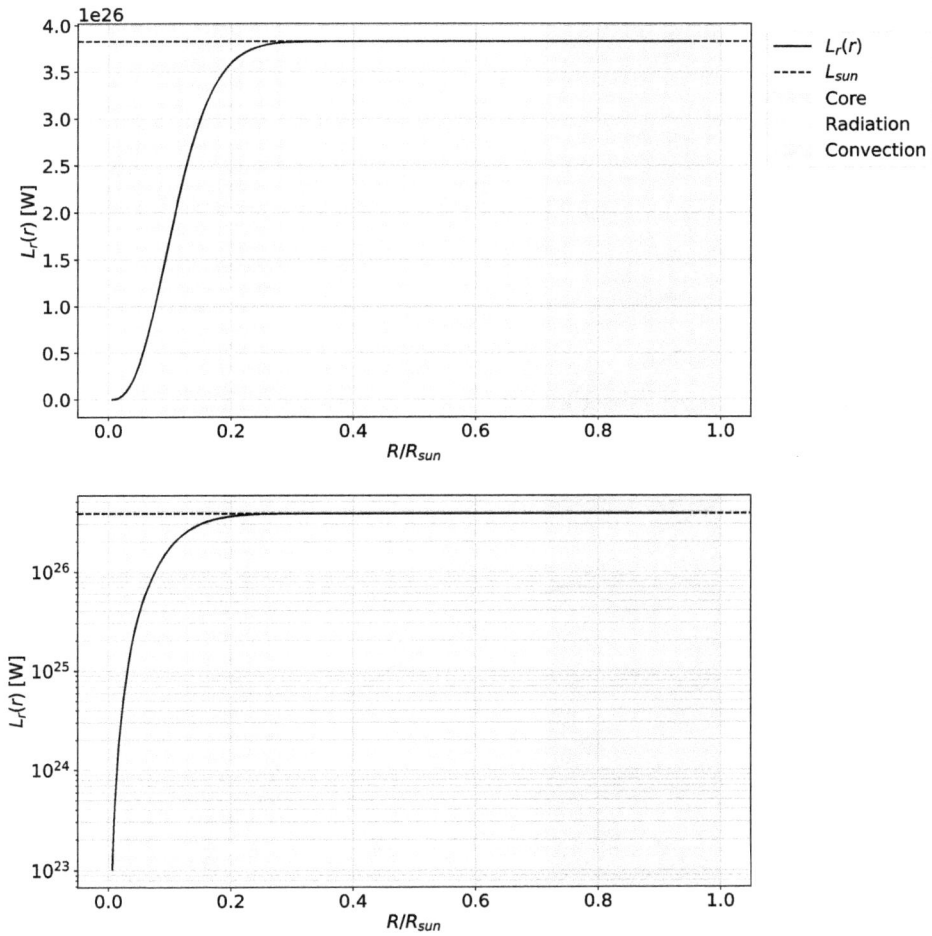

Figure 23.5. Enclosed luminosity as a function of radius, the solution to Equation (23.22). Note that all of the energy generation happens in the core; outside the core, the enclosed luminosity remains constant and simply diffuses to the surface.

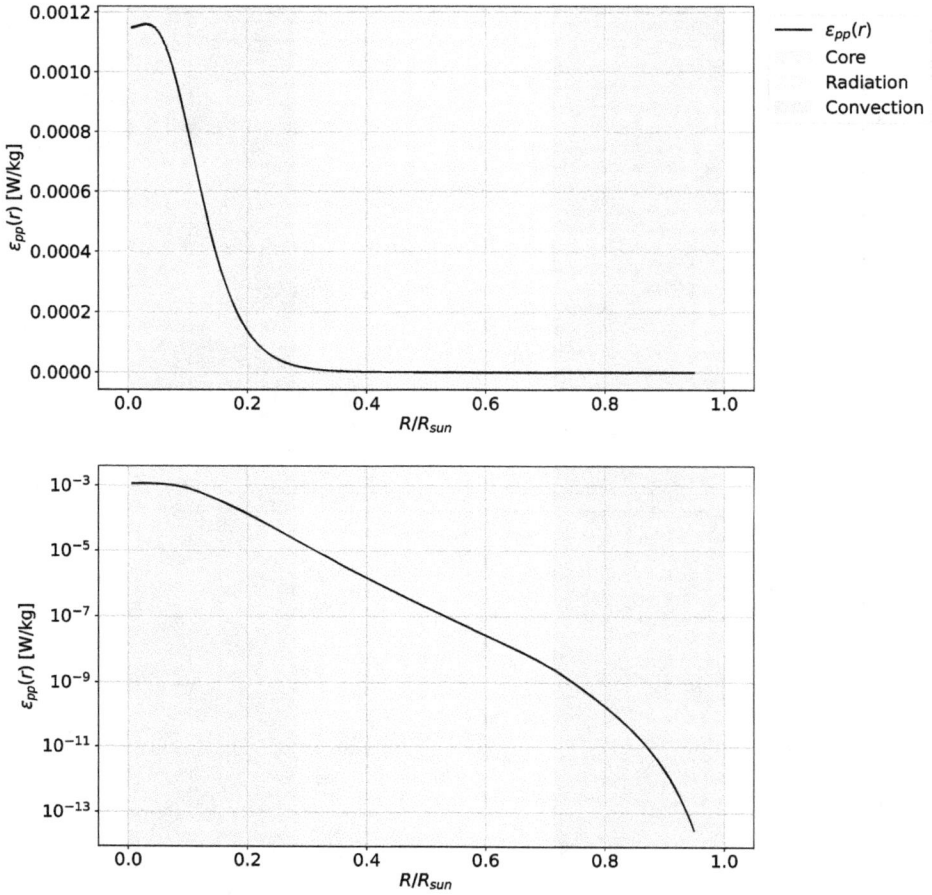

Figure 23.6. Nuclear energy generation as a function of radius, computed from Equation (23.5). Note that this equation never has the energy generation drop entirely to zero, though it is at ~1% of the central value by the edge of the core. This numerical quirk is not a feature of the more detailed model Bahcall et al. (2001); see Figure 23.13.

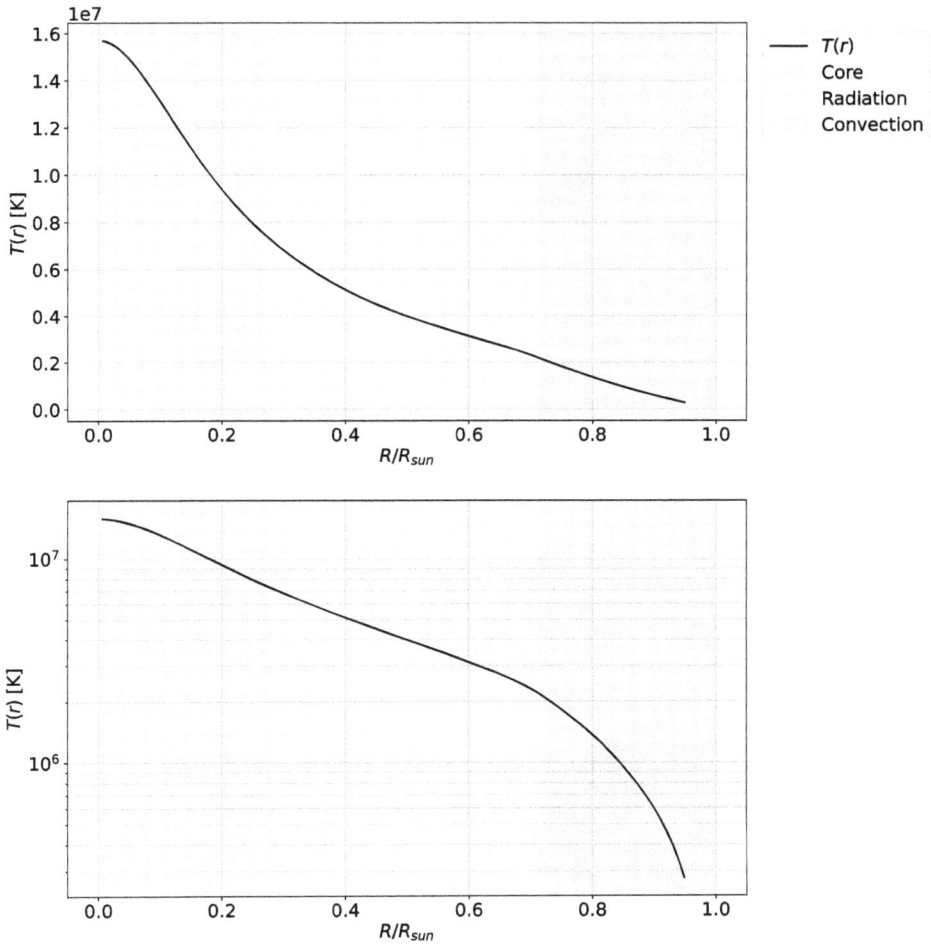

Figure 23.7. Temperature as a function of radius, the solution to Equation (23.23). Note that the temperature changes less dramatically, in a fractional sense, that the other structure variables. Note too, though, that the since the model does not go all the way to the full solar radius, we do not actually reach the surface temperature in these plots, but are about an order of magnitude higher at the largest radius plotted.

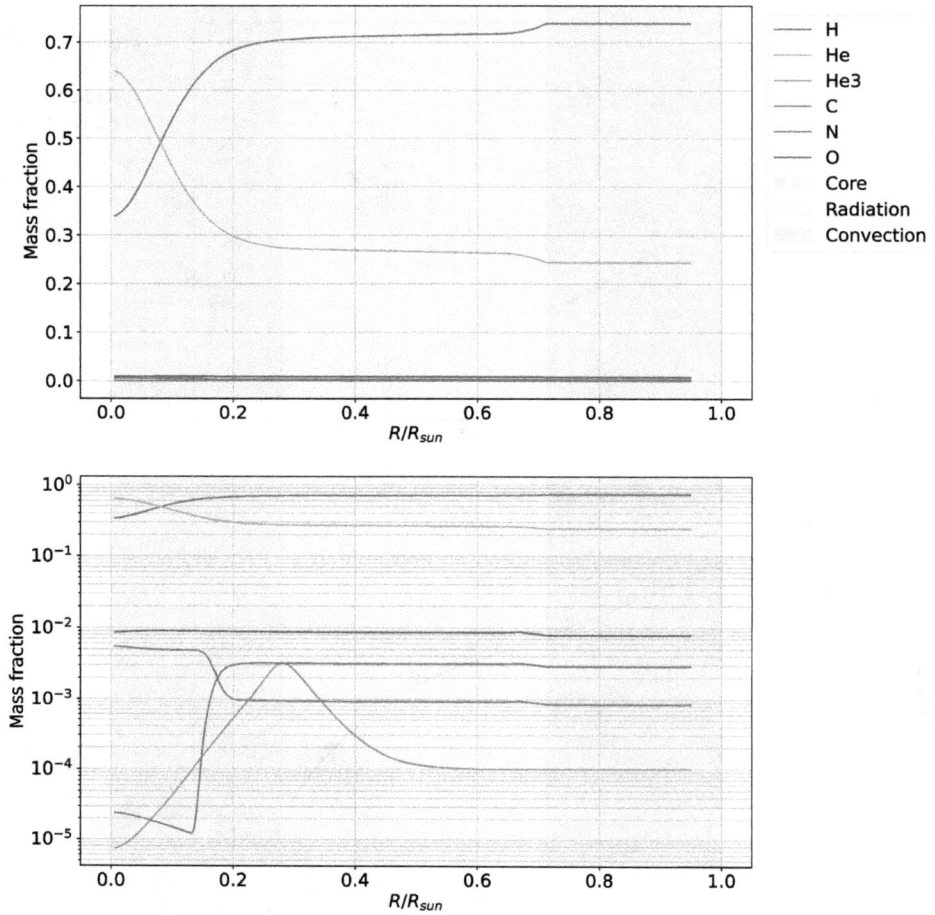

Figure 23.8. Composition as a function of radius. Note that only the most abundant elements are tracked, and the key feature is the conversion of hydrogen into helium in the core. Note, however, the interesting changes in abundance of He3 in particular at the core boundary, and all of the elements at the radiative to convective transition.

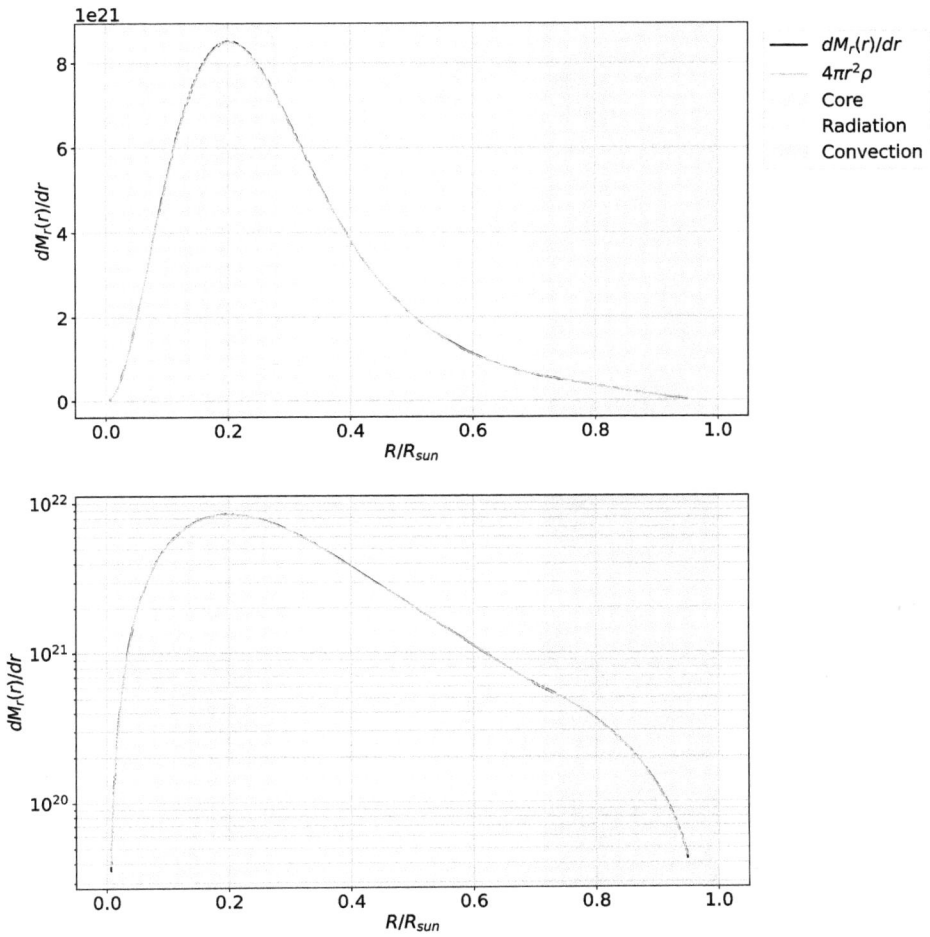

Figure 23.9. Comparing the numerical derivative of the model M_r (black) versus the right-hand size of Equation (23.21) (orange).

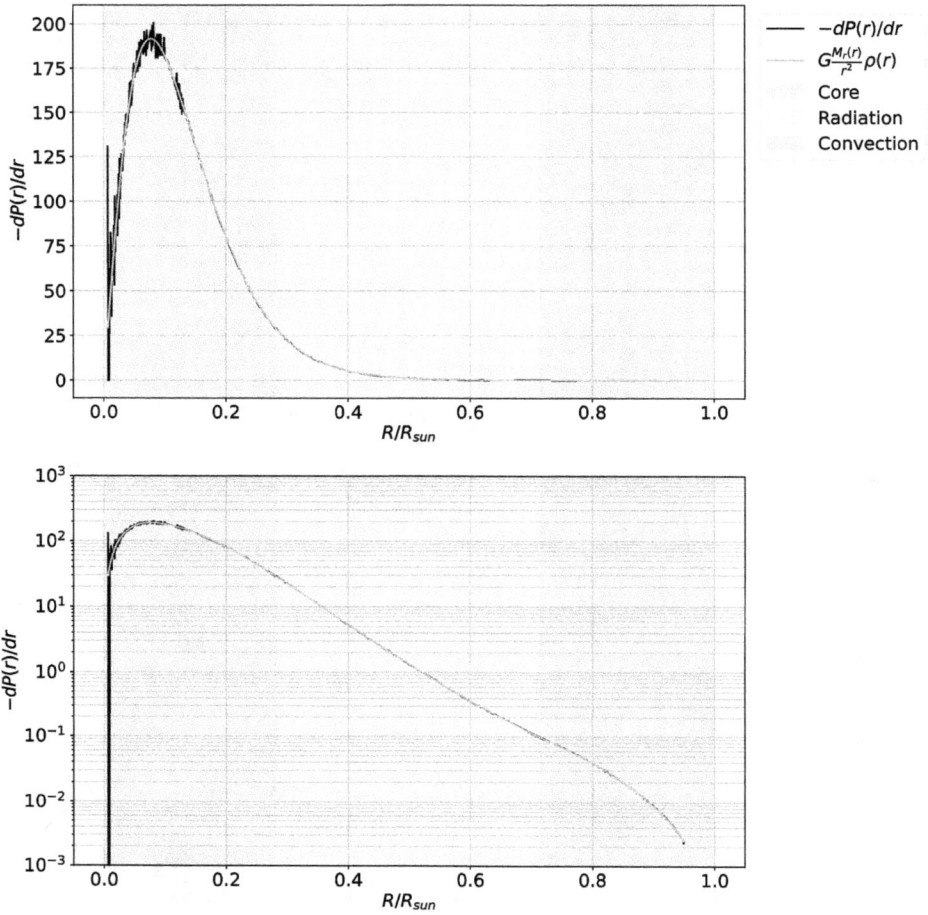

Figure 23.10. Comparing the numerical derivative of the model P (black) versus the right-hand size of Equation (23.20) (orange).

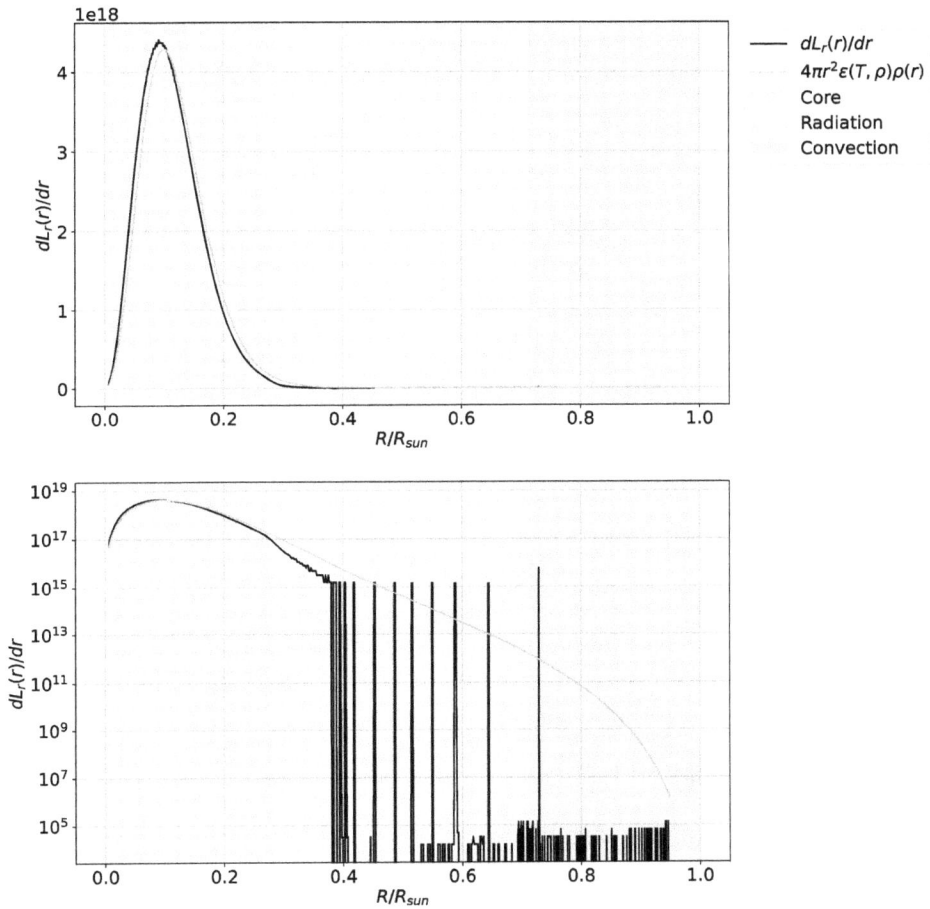

Figure 23.11. Comparing the numerical derivative of the model L_r (black) versus the right-hand size of Equation (23.22) (orange). Here we compare our simple model of the nuclear energy generation to the more sophisticated model, which correctly implements the Coulomb barrier between nuclei shutting off fusion outside the high temperatures and densities of the core.

Reference

Bahcall, J. N., Pinsonneault, M. H., & Basu, S. 2001, ApJ, 555, 990

An Introduction to Astrophysics with Python
Stars and planets
James Aguirre

Appendix A

A "Derivation" of the Planck Blackbody Formula

We provide a simplified derivation of the Planck blackbody formula using a combination of thermodynamic and simplified quantum mechanical arguments with a two-state system. This allows us to define the Einstein A parameter, which is related to the damping coefficient from Chapter 22. This provides us with a motivation for the functional forms presented in Chapter 14.

Here we present a heuristic derivation of the form of I_ν in the case that photons of frequency ν are in thermal equilibrium with matter with an internal energy difference $\Delta E = h\nu$. We will assume that such an energy difference exists in the matter for every possible ν we might be interested in.

Consider just two energy levels in the piece of matter: to describe them, we need to specify the *degeneracies* of the two states, and their energy difference. We use the Boltzmann equation (Equation (13.1)) to specify the relative numbers of atoms in the two states, described by the quantities n_1 and n_2:

$$\boxed{\frac{n_2}{n_1} = \frac{g_2}{g_1} \exp\left(-\frac{\Delta E}{kT}\right) = \frac{g_2}{g_1} \exp\left(-\frac{(E_2 - E_1)}{kT}\right)} \tag{A.1}$$

To characterize the rate at which atoms in the 2 state are going to the 1 state we write

$$\frac{dn_2}{dt} = -n_2 B_{21} J_\nu - n_2 A_{21} \tag{A.2}$$

This process involves both *spontaneous* and *stimulated* emission. A_{21} is the so-called Einstein coefficient for the spontaneous emission of a photon. In discussing

doi:10.1088/2514-3433/ade5f6ch24

the natural width of the lines in Chapter 22, we introduce the "damping constant" γ_n (Equation (22.1)), which is related to the uncertainty of the natural lifetime of the excited state. A more precise definition of the damping constant takes into account all of the possible transitions from that excited state, so that

$$\gamma_n = \sum_{n' < n} A_{nn'} \tag{A.3}$$

where $A_{nn'}$ is the *Einstein coefficient for spontaneous emission from state n to n'*. $A_{nn'}$ has units of Hz, and is related to the lifetime uncertainty between states n and n'.

B_{21} describes the *stimulated emission* from $2 \rightarrow 1$, which depends on the incident photon flux at frequency ν, described by J_ν. Its units are more complicated:

$$[B_{21}] = \frac{\text{m}^2\text{Hz}}{\text{Ws}} = \frac{\text{m}^2}{\text{Js}} \tag{A.4}$$

Similarly, rate of change of atoms in the first state can be written as

$$\frac{dn_1}{dt} = -n_1 B_{12} J_\nu \tag{A.5}$$

where B_{12} describes absorption. In a steady state, the number of atoms going from $2 \rightarrow 1$ must equal the number going from $1 \rightarrow 2$, which means

$$n_1 B_{12} J_\nu = n_2 B_{21} J_\nu + n_2 A_{21} \tag{A.6}$$

or, simply rearranging

$$\frac{n_2}{n_1} = \frac{B_{12} J_\nu}{B_{21} J_\nu + A_{21}} \tag{A.7}$$

Now we can use the Boltzmann equation to replace the left-hand side:

$$\frac{n_2}{n_1} = \frac{B_{12} J_\nu}{B_{21} J_\nu + A_{21}} = \frac{g_2}{g_1} \exp\left(-\frac{\Delta E}{kT}\right) \tag{A.8}$$

In the limit $T \rightarrow \infty$, the radiation field is strong, and stimulated emission dominates over spontaneous emission, or $B_{21} J_\nu \gg A_{21}$ and we find that there must be a relation between the B's:

$$g_1 B_{12} = g_2 B_{21} \tag{A.9}$$

Plugging this back in leads to

$$\frac{g_2 B_{21} J_\nu}{B_{21} J_\nu + A_{21}} = g_2 \exp\left(-\frac{\Delta E}{kT}\right) \tag{A.10}$$

$$B_{21} J_\nu = \exp\left(-\frac{\Delta E}{kT}\right)(B_{21} J_\nu + A_{21}) \tag{A.11}$$

$$B_{21} J_\nu\left(1 - \exp\left(-\frac{\Delta E}{kT}\right)\right) = A_{21} \exp\left(-\frac{\Delta E}{kT}\right) \tag{A.12}$$

and

$$J_\nu = \frac{A_{21}}{B_{21}}\left[\exp\left(\frac{\Delta E}{kT}\right) - 1\right]^{-1} \tag{A.13}$$

To get further, we can invoke some properties of the quantum mechanics of radiation, or we can use some empirical relations (which were already known by the beginning of the 20th century). We'll go with the latter, but we note that these relations can be gotten from theory.

Empirically, in the high-energy limit $h\nu/kT \gg 1$, the form of the blackbody spectrum is given

$$J_\nu \propto \nu^3 \exp(-h\nu/kT) \tag{A.14}$$

Comparing, this allows us to conclude

$$\frac{A_{21}}{B_{21}} \propto \nu^3 \tag{A.15}$$

and

$$\Delta E = h\nu \tag{A.16}$$

To find the constant of proportionality, we need to look at the opposite limit, $h\nu/kT \ll 1$, where again empirically

$$J_\nu = \frac{2\nu^2}{c^2}kT \tag{A.17}$$

In this limit, we have

$$\frac{A_{21}}{B_{21}}\frac{kT}{h\nu} = \frac{2\nu^2}{c^2}kT \tag{A.18}$$

Thus

$$\frac{A_{21}}{B_{21}} = \frac{2h\nu^3}{c^2} \tag{A.19}$$

and finally

$$J_\nu = \frac{2h\nu^3}{c^2}\left[\exp\left(\frac{h\nu}{kT}\right) - 1\right]^{-1} \tag{A.20}$$

Looking back, we can see that if I_ν does not depend on angle—in other words, is isotropic—that

$$J_\nu = \frac{1}{4\pi}\int I_\nu d\Omega = \frac{I_\nu}{4\pi}\int d\Omega = \frac{I_\nu}{4\pi}4\pi = I_\nu \tag{A.21}$$

The *Planck formula for blackbody radiation* is:

$$B_\nu = \frac{2h\nu^3}{c^2}\left[\exp\left(\frac{h\nu}{kT}\right) - 1\right]^{-1} \tag{A.22}$$

We can transform B_ν to B_λ by considering the relation

$$B_\lambda = B_\nu \left|\frac{d\nu}{d\lambda}\right| \tag{A.23}$$

In the function B_ν we substitute $\nu = c/\lambda$ to change the frequency dependence to wavelength dependence, and then we need

$$\frac{d\nu}{d\lambda} = -\frac{c}{\lambda^2} \tag{A.24}$$

This leads to

$$B_\lambda = \frac{2hc^2}{\lambda^5}\frac{1}{\exp\dfrac{hc}{\lambda kT} - 1} \tag{A.25}$$

www.ingramcontent.com/pod-product-compliance
Lightning Source LLC
Chambersburg PA
CBHW080523220326
41599CB00032B/6176